Interdisciplinary Applied Mathematics

Volume 34

Editors
S.S. Antman **J.E. Marsden**
L. Sirovich **S. Wiggins**

Geophysics and Planetary Sciences

Imaging, Vision, and Graphics
D. Geman

Mathematical Biology
L. Glass, J.D. Murray

Mechanics and Materials
R.V. Kohn

Systems and Control
S.S. Sastry, P.S. Krishnaprasad

Problems in engineering, computational science, and the physical and biological sciences are using increasingly sophisticated mathematical techniques. Thus, the bridge between the mathematical sciences and other disciplines is heavily traveled. The correspondingly increased dialog between the disciplines has led to the establishment of the series: *Interdisciplinary Applied Mathematics*.

The purpose of this series is to meet the current and future needs for the interaction between various science and technology areas on the one hand and mathematics on the other. This is done, firstly, by encouraging the ways that mathematics may be applied in traditional areas, as well as point towards new and innovative areas of applications; and, secondly, by encouraging other scientific disciplines to engage in a dialog with mathematicians outlining their problems to both access new methods and suggest innovative developments within mathematics itself.

The series will consist of monographs and high-level texts from researchers working on the interplay between mathematics and other fields of science and technology.

Interdisciplinary Applied Mathematics

Agnès Desolneux   Lionel Moisan
Jean-Michel Morel

From Gestalt Theory to Image Analysis

A Probabilistic Approach

A. Desolneux
Université Paris Descartes
MAP5 (CNRS UMR 8145)
45, rue des Saints-Pères
75270 Paris cedex 06, France
desolneux@math-info.univ-paris5.fr

L. Moisan
Université Paris Descartes
MAP5 (CNRS UMR 8145)
45, rue des Saints-Pères
75270 Paris cedex 06, France
moisan@math-info.univ-paris5.fr

J.-M. Morel
Ecole Normale Supérieure de Cachan, CMLA
61, av. du Président Wilson
94235 Cachan Cédex
France
Jean-Michel.Morel@cmla.ens-cachan.fr

Editors

S.S. Antman
Department of Mathematics
and
Institute for Physical Science
and Technology
University of Maryland
College Park, MD 20742, USA
ssa@math.umd.edu

J.E. Marsden
Control and Dynamical Systems
Mail Code 107-81
California Institute of Technology
Pasadena, CA 91125, USA
marsden@cds.caltech.edu

L. Sirovich
Division of Applied Mathematics
Brown University
Providence, RI 02912, USA
chico@camelot.mssm.edu

S. Wiggins
School of Mathematics
University of Bristol
Bristol BS8 1TW, UK
s.wiggins@bris.ac.uk

ISBN: 978-1-4419-2481-0   e-ISBN: 978-0-387-74378-3
DOI: 10.1007/978-0-387-74378-3

Mathematics Subject Classification (2000): 62H35, 68T45, 68U10

Printed on acid-free paper

9 8 7 6 5 4 3 2 1

springer.com

Preface

The theory in these notes was taught between 2002 and 2005 at the graduate schools of Ecole Normale Supérieure de Cachan, Ecole Polytechnique de Palaiseau, Universitat Pompeu Fabra, Barcelona, Universitat de les Illes of Balears, Palma, and University of California at Los Angeles. It is also being taught by Andrès Almansa at the Facultad de Ingeneria, Montevideo.

This text will be of interest to several kinds of audience. Our teaching experience proves that specialists in image analysis and computer vision find the text easy at the computer vision side and accessible on the mathematical level. The prerequisites are elementary calculus and probability from the first two undergraduate years of any science course. All slightly more advanced notions in probability (inequalities, stochastic geometry, large deviations, etc.) will be either proved in the text or detailed in several exercises at the end of each chapter. We have always asked the students to do all exercises and they usually succeed regardless of what their science background is. The mathematics students do not find the mathematics difficult and easily learn through the text itself what is needed in vision psychology and the practice of computer vision. The text aims at being self-contained in all three aspects: mathematics, vision, and algorithms. We will in particular explain what a digital image is and how the elementary structures can be computed.

We wish to emphasize why we are publishing these notes in a mathematics collection. The main question treated in this course is the visual perception of geometric structure. We hope this is a theme of interest for all mathematicians and all the more if visual perception can receive –up to a certain limit we cannot yet fix– a fully mathematical treatment. In these lectures, we rely on only four formal principles, each one taken from perception theory, but receiving here a simple mathematical definition. These mathematically elementary principles are the *Shannon-Nyquist principle*, the *contrast invariance* principle, *the isotropy principle* and the *Helmholtz principle*. The first three principles are classical and easily understood. We will just state them along with their straightforward consequences. Thus, the text is mainly dedicated to one principle, the Helmholtz principle. Informally, it states that *there is no perception in white noise*. A white noise image is an image whose samples are identically distributed independent random variables. The view of a white sheet of paper in daylight gives a fair idea of what white noise is. The whole work will be to draw from this impossibility of seing something on a white sheet a series of mathematical techniques and algorithms analyzing digital images and "seeing" the geometric structures they contain.

Most experiments are performed on digital every-day photographs, as they present a variety of geometric structures that exceeds by far any mathematical modeling and are therefore apt for checking any generic image analysis algorithm. A warning to mathematicians: It would be fallacious to deduce from the above lines that we are proposing a definition of geometric structure for all real functions. Such a definition would include all geometries invented by mathematicians. Now, the mathematician's real functions are, from the physical or perceptual viewpoint, impossible objects with infinite resolution and that therefore have infinite details and structures on all scales. Digital *signals*, or *images*, are surely functions, but with the essential limitation of having a finite resolution permitting a finite sampling (they are band-limited, by the Shannon-Nyquist principle). Thus, in order to deal with digital images, a mathematician has to abandon the infinite resolution paradise and step into a finite world where geometric structures must all the same be found and proven. They can even be found with an almost infinite degree of certainty; how sure we are of them is precisely what this book is about.

The authors are indebted to their collaborators for their many comments and corrections, and more particularly to Andrès Almansa, Jérémie Jakubowicz, Gary Hewer, Carol Hewer, and Nick Chriss. Most of the algorithms used for the experiments are implemented in the public software MegaWave. The research that led to the development of the present theory was mainly developed at the University Paris-Dauphine (Ceremade) and at the Centre de Mathématiques et Leurs Applications, ENS Cachan and CNRS. It was partially financed during the past 6 years by the Centre National d'Etudes Spatiales, the Office of Naval Research, and NICOP under grant N00014-97-1-0839 and the Fondation les Treilles. We thank very much Bernard Rougé, Dick Lau, Wen Masters, Reza Malek-Madani, and James Greenberg for their interest and constant support. The authors are grateful to Jean Bretagnolle, Nicolas Vayatis, Frédéric Guichard, Isabelle Gaudron-Trouvé, and Guillermo Sapiro for valuable suggestions and comments.

Contents

Chapter 1
Introduction

1.1 Gestalt Theory and Computer Vision

Why do we interpret stimuli arriving at our retina as straight lines, squares, circles, and any kind of other familiar shape? This question may look incongruous: What is more natural than recognizing a "straight line" in a straight line image, a "blue cube" in a blue cube image? When we believe we see a straight line, the actual stimulus on our retina does not have much to do with the mathematical representation of a continuous, infinitely thin, and straight stroke. All images, as rough data, are a pointillist datum made of more or less dark or colored dots corresponding to local retina cell stimuli. This total lack of structure is equally true for digital images made of *pixels*, namely square colored dots of a fixed size.

How groups of those pixels are built into spatially extended visual objects is, as Gaetano Kanizsa [Kan97] called it, one of the major "enigmas of perception." The enigma consists of the identification performed between a certain subgroup of the perceptum (here the rough datum on the retina) and some physical object, or even some geometric abstraction like a straight line. Such identification must obey general laws and principles, which we will call *principles of visual reconstruction* (this term is borrowed from Gombrich [Gom71]).

There is, to the best of our knowledge, a single substantial scientific attempt to state the laws of visual reconstruction: the Gestalt Theory. The program of this school is first given in Max Wertheimer's 1923 founding paper [Wer23]. In the Wertheimer program there are two kinds of organizing laws. The first kind are grouping laws, which, starting from the atomic local level, recursively construct larger groups in the perceived image. Each grouping law focuses on a single quality (color, shape, direction...). The second kind are principles governing the collaboration and conflicts of gestalt laws. In its 1975 last edition, the gestalt "Bible" *Gesetze des Sehens*, Wolfgang Metzger [Met75] gave a broad overview of the results of 50 years of research. It yielded an extensive classification of grouping laws and many insights about more general gestalt principles governing the interaction (collaboration and conflicts) of grouping laws. These results rely on an incredibly rich and imaginative collection of test figures demonstrating those laws.

1

At about the same time Metzger's book was published, computer vision was an emerging new discipline at the meeting point of artificial intelligence and robotics. Although the foundation of signal sampling theory by Claude Shannon [Sha48] was already 20 years old, computers were able to deal with images with some efficiency only at the beginning of the seventies. Two things are noticeable:

- Computer Vision did not at first use the Gestalt Theory results: David Marr's [Mar82] founding book involves much more neurophysiology than phenomenology. Also, its program and the robotics program [Hor87] founded their hopes on binocular stereo vision. This was in contradiction with the results explained at length in many of Metzger's chapters dedicated to *Tiefensehen* (depth perception). These chapters demonstrate that binocular stereo vision is a *parent pauvre* in human depth perception.
- Conversely, Shannon's information theory does not seem to have influenced gestalt research as far as we can judge from Kanizsa's and Metzger's books. Gestalt Theory does not take into account the finite sampled structure of digital images! The only brilliant exception is Attneave's attempt [Att54] to adapt sampling theory to shape perception.

This lack of initial interaction is surprising. Both disciplines have attempted to answer the following question: how to arrive at global percepts — be they visual objects or gestalts — from the local, atomic information contained in an image?

In these notes, we tentatively translate the Wertheimer program into a mathematics and computer vision program. This translation is not straightforward, since Gestalt Theory did not address two fundamental matters: image sampling and image information measurements. Using them, we will be able to translate qualitative geometric phenomenological observations into quantitative laws and eventually to numerical simulations of gestalt grouping laws.

One can distinguish at first two kinds of laws in Gestalt Theory:

- practical grouping laws (like vicinity or similarity), whose aim it is to build up *partial gestalts*, namely elementary perception building blocks;
- gestalt principles like masking or *articulazione senza resti*, whose aim it is to operate a synthesis between the partial groups obtained by elementary grouping laws.

See Figure 1.1 for a first example of these gestalt laws. Not surprisingly, phenomenology-styled gestalt principles have no direct mathematical translation. Actually, several mathematical principles were probably too straightforward to be stated by psychologists. Yet, a mathematical analysis cannot leave them in the dark. For instance, no translation invariance principle is proposed in Gestalt Theory, in contrast with signal and image analysis, where it takes a central role. Gestaltists ignored the mathematical definition of digital image and never used resolution (for example) as a precise concept. Most of their grouping laws and principles, although having an obvious mathematical meaning, remained imprecise. Several of the main issues in digital image analysis, namely the role of noise and blur in image formation, were not quantitatively and even not qualitatively considered.

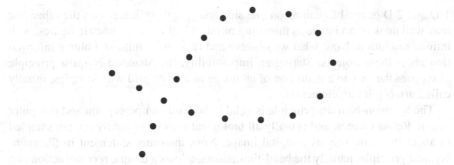

Fig. 1.1 A first example of the two kinds of gestalt laws mentioned. Black dots are grouped together according to elementary grouping laws like vicinity, similarity of shape, similarity of color, and good continuation. These dots form a loop-like curve and not a closed curve plus two small remaining curves: This is an illustration of the global gestalt principle of *articulazione senza resti*.

1.2 Basic Principles of Computer Vision

A principle is merely a statement of an impossibility (A. Koyré). A few principles lead to quantitative laws in mechanics; their role has to be the same in computer vision. Of course, all computer vision algorithms deriving from principles should be free of parameters left to the user. This requirement may look straightforward but is not acknowledged in the Computer Vision literature. Leaving parameters to the user's choice means that something escaped from the modeling — in general, a hidden principle.

As we mentioned earlier, the main body of these lectures is dedicated to the thorough study of the consequences of Helmholtz's principle, which, as far as we know, receives its first mathematical systematic study here. The other three basic and well-known principles are the *Shannon sampling principle*, defining digital images and fixing a bound to the amount of information contained in them, the *Wertheimer contrast invariance principle*, which forbids taking literally the actual values of gray levels, and the *isotropy* principle, which requires image analysis to be invariant with respect to translations and rotations.

In physics, principles can lead to quantitative laws and very exact predictions based on formal or numerical calculations. In Computer Vision, our aim is to predict all basic perceptions associated with a digital image. These predictions must be based on parameter-free algorithms (i.e., algorithms that can be run on any digital image without human intervention).

We start with an analysis of the three basic principles and explain why they yield image processing algorithms.

Principle 1 (Shannon-Nyquist, definition of signals and images) *Any image or signal, including noisy signals, is a band-limited function sampled on a bounded, periodic grid.*

This principle says first that we cannot hope for an infinite resolution or an infinite amount of information in a digital image. This makes a big difference between

1-D and 2-D general functions on one side and signals or images on the other. We may well think of an image as mirroring physical bodies, or geometric figures, with infinite resolution. Now, what we observe and register is finite and blurry information about these objects. Stating an impossibility, the Shannon-Nyquist principle also opens the way to a definition of an image as a finite grid with samples, usually called *pixels* (picture elements).

The Shannon-Nyquist principle is valid in both human perception and computer vision. Retina images, and actually all biological eyes from the fly up, are sampled in about the same way as a digital image. Now, the other statement in Shannon-Nyquist principle, namely the band-limitedness, allows a unique reconstruction of a continuous image from its samples. If that principle is not respected, the interpolated image is not invariant with respect to the sampling grid and aliasing artifacts appear, as pointed out in Figure 1.2.

Algorithm 1 *Let $u(x,y)$ be a real function on the plane and $\hat{u}$ its Fourier transform. If $\text{Support}(\hat{u}) \subset [-\pi,\pi]^2$, then u can be reconstructed from the samples $u(m,n)$ by*

$$u(x,y) = \sum_{(m,n)\in\mathbb{Z}^2} u(m,n) \frac{\sin\left(\pi(x-m)\right)}{\pi(x-m)} \frac{\sin\left(\pi(y-n)\right)}{\pi(y-n)}$$

In practice, only a finite number of samples $u(m,n)$ can be observed. Thus, by the above formula, digital images turn out to be trigonometric polynomials.

Since it must be sampled, every image has a critical resolution: twice the distance between two pixels. This mesh will be used thoroughly in these notes. Consequently,

Fig. 1.2 On the left, a well-sampled image according to the Shannon-Nyquist principle. The relations between sample distances and the Fourier spectrum content of the image are in conformity with Principle 1 and Algorithm 1. If these conditions are not respected, the image may undergo severe distortions, as shown on the right.

there is a universal image format, namely a (usually square or rectangular) grid of "pixels". Since the gray level at each point is also quantized and bounded, all images have a finite maximum amount of information, namely the number of points in the sampling grid (the so-called pixels = picture elements) multiplied by roughly 8 bits/pixel (gray level) or 24 bits in case of color images. In other terms, the gray level and each color is encoded by an integer ranging from 0 to 255.

Principle 2 (Wertheimer's contrast invariance principle) *Image interpretation does not depend on actual values of the gray levels, but only on their relative values.*

Again, this principle states an impossibility, namely the impossibility of taking digital images as reliable physical measurements of the illumination and reflectance materials of the photographed objects. On the positive side, it tells us where to look to get reliable information. We can rely on information that only depends on the order of gray levels — that is to say, *contrast invariant* information.

The Wertheimer principle was applied in Computer Vision by Matheron and Serra [Ser82], who noticed that upper or lower level sets and the level lines of an image contain the shape information, independently of contrast information. Also, because of the same principle, we will only retain the gradient orientation and not the modulus of gradient as relevant information in images. For Matheron and Serra, the building blocks for image analysis are given, for example, by the upper level sets. As usual with a good principle, one gets a good simple algorithm. Wertheimer's principle yields the basic algorithm of mathematical morphology : it parses an image into a set of sets, the upper level sets. These sets can be used for many tasks, including shape analysis.

Algorithm 2 *Let $u(x,y)$ be a gray-level image. The upper level sets of u are defined by*

$$\chi_\lambda(u) = \{(x,y), u(x,y) \geq \lambda\}.$$

The set of all level sets $\{\chi_\lambda, \lambda \in \mathbb{R}\}$ is contrast invariant and u can be reconstructed from its level sets by

$$u(x,y) = \sup\{\lambda, (x,y) \in \chi_\lambda(u)\}.$$

A still better representation is obtained by encoding an image as the set of its level lines, the level lines being defined as the boundaries of level sets. The interpolated digital image being smooth by the Shannon-Nyquist principle, the level lines are Jordan curves for almost every level (see Figure 1.3).

Principle 3 (Helmholtz principle, first stated by D. Lowe [Low85]) *Gestalts are sets of points whose (geometric regular) spatial arrangement could not occur in noise.*

This statement is a bit vague. It is the aim of the present notes to formalize it. As we will prove in detail with geometric probability arguments, this principle yields

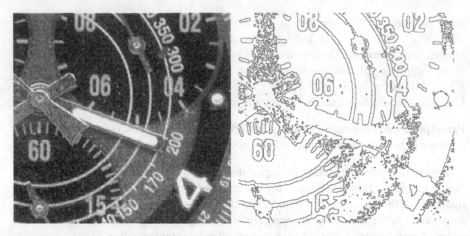

Fig. 1.3 Contrast invariant features deriving from Wertheimer's principle: On the right, some image level lines, or isophotes, corresponding to the gray level $\lambda = 128$. According to Wertheimer's principle, the level lines contain the whole shape information.

algorithms for all grouping laws and therefore permits us to compute what we will call "partial gestalts". A weaker form of this principle can be stated as "there is no perceptual structure in white noise".

In other terms, every structure that shows too much geometric regularity to be found by chance in noise calls attention and becomes a perception. The Helmholtz principle is at work in Dostoievsky's *The Player*, where specific sequences of black or red are noticed by the players as exceptional, or meaningful, at roulette: If a sequence of 20 consecutive "red" occurs, this is considered noticeable. Yet, all other possible red and black sequences of the same length have the same probability. Most of them occur without raising interest: Only those corresponding to a "grouping law" — here the color constancy — impress the observer. We will analyze with much detail this example and other ones in Chapter 3. The detection of alignments in a digital image is very close to the Dostoievsky example.

An alignment in a digital image is defined as a large enough set of sample points on a line segment at which the image gradient is orthogonal enough to the segment to make this coincidence unlikely in a white noise image.

The algorithm to follow is, as we will prove, a direct consequence of the three basic principles, namely the Shannon-Nyquist interpolation and sampling principle, Wertheimer's contrast invariance principle, and the Helmholtz grouping principle. It summarizes the theory we will develop in Chapters 5 and 6.

Algorithm 3 (Computing Alignments)

– *Let N_S be the number of segments joining pixels of the image.*
– *Let $0 \leq p \leq 1$ be an angular precision (arbitrary).*
– *Let S be a segment with length l and with k sample points aligned at precision p.*

Fig. 1.4 Left: original aerial view (source: INRIA); middle: maximal meaningful alignments; right: maximal meaningful boundaries.

- *Then the number of false alarms of this event in a noise Shannon image of the same size is*

$$\mathrm{NFA}(l,k,p) = N_S \sum_{j=k}^{l} \binom{l}{j} p^j (1-p)^{l-j}.$$

- *An alignment is meaningful if* $\mathrm{NFA}(l,k,p) \leq 1$.

We will apply exactly the same principles to derive a definition of "perceptual boundaries" and an unsupervised algorithm computing them in a digital image. The next informal definition will be made rigorous in Chapter 9.

A perceptual boundary is defined as a level line whose points have a "large enough" gradient, so that no such line is likely to occur in a white noise with the same overall contrast.

Figure 1.4 shows meaningful alignments and meaningful boundaries detected according to the preceding definitions. The notion of "maximal meaningfulness" will be developed in Chapter 6. In addition to the Helmholtz principle, Figure 1.4 and all experiments in the book will extensively use the *exclusion principle*, presented in Chapter 6. Roughly speaking, this principle forbids a visual object to belong to two different groups that have been built by the same grouping law. This implies, for example, that two different alignments, or boundaries, cannot overlap. Here is our plan.

- Chapter 1 is the present short introduction.
- Chapter 2 is dedicated to a critical description of gestalt grouping laws and gestalt principles.
- Chapter 3 states and formalizes the Helmholtz principle by discussing several examples, including the recognition of simple shapes, Dostoievsky's roulette, and alignments in a image made of dots.
- Chapter 4 gives estimates of the central function in the whole book, the so-called "number of false alarms" (NFA), which in most cases can be computed as a tail of a binomial law.

– Chapter 5 defines "meaningful alignments" in a digital image and their *number of false alarms* as a function of three (observed) parameters, namely precision, length of the alignment, and number of aligned points. This is somehow the central chapter, as all other detections can be viewed as variants of the alignment detection.
– Chapter 6 is an introduction to the exclusion principle, followed by a definition of "maximal meaningful" gestalts. In continuation, it is proven that maximal meaningful alignments do not overlap and therefore obey the exclusion principle.
– Chapter 7 treats the most basic grouping task: how to group objects that turn out to have one quality in common, be it color, orientation, size, or other qualities. Again, "meaningful groups" are defined and it is again proved that maximal meaningful groups do not overlap.
– Chapter 8 treats the detection of one of the relevant geometric structures in painting, also essential in photogrammetry: the vanishing points. They are defined as points at which exceptionally many alignments meet. This is a "second-order" gestalt.
– Chapter 9 extends the theory to one of the most controversial detection problems in image analysis, the so-called segmentation, or edge detection theory. All state-of-the art methods depend on several user's parameters (usually two or more). A tentative definition of meaningful contours by the Helmholtz principle eliminates all the parameters.
– Chapter 10 compares the new theory with the state-of-art theories, in particular with the "active contours" or "snakes" theory. A very direct link of "meaningful boundaries" to "snakes" is established.
– Chapter 11 proposes a theory to compute, by the Helmholtz principle, clusters in an image made of dots. This is the classical *vicinity* gestalt: Objects are grouped just because they are closer to each other than to any other object.
– Chapter 12 addresses a key problem of photogrammetry: the binocular stereo vision. Digital binocular vision is based on the detection of special points like corners in both images. These points are grouped by pairs by computer vision algorithms. If the groups are right, the pairs of points define an *epipolar geometry* permitting one to build a line-to-line mapping from one image onto the other one. The main problem turns out to be, in practice, the large number of wrong pairs. Using the Helmholtz principle permits us to detect the right and more precise pairs of points and therefore to reconstruct the epipolar geometry of the pair of images.
– Chapter 13 describes two simple psychophysical experiments to check whether the *perception thresholds* match the ones predicted by the Helmholtz principle. One of the experiments deals with the detection of squares in a noisy environment and the other one deals with alignment detection.
– Chapter 14 presents a synopsis of results with a table of formulas for all gestalts. It also discusses some experiments showing how gestalt detectors could "collaborate". This chapter ends with a list of unsolved questions and puzzling experiments showing the limits in the application of the found principles. In particular,

the notion of "conflict" between gestalts, raised by gestaltists, has no satisfactory formal answer so far.
- Chapter 15 discusses precursory and alternative theories. It also contains sections about the relation between the Number of False Alarms and the classical statistical framework of hypothesis testing. It ends with a discussion about Bayesian framework and the Minimum Description Length principle.

Chapter 2
Gestalt Theory

In this chapter, we start in Section 2.1 with some examples of optic-geometric illusions and then give, in Section 2.2, an account of Gestalt Theory, centered on the initial 1923 Wertheimer program. In Section 2.3 the focus is on the problems raised by the synthesis of groups obtained by partial grouping laws. Following Kanizsa, we will address the conflicts between these laws and the masking phenomenon. In Section 2.4 several quantitative aspects implicit in Kanizsa's definition of masking are indicated. It is shown that one particular kind of masking, Kanizsa's *masking by texture*, may lead to a computational procedure.

2.1 Before Gestaltism: Optic-Geometric Illusions

Naturally enough, the study of vision started with a careful examination by physicists and biologists of the eye, thought of as an optical apparatus. Two of the most complete theories come from Helmholtz [vH99] and Hering [Her20]. This analysis naturally led to checking how reliably visual percepts related to the physical objects. This led to the discovery of several now-famous aberrations. We will not explain them all, but just those that are closer to our subject, namely the geometric aberrations, usually called optic-geometric illusions. They consist of figures with simple geometric arrangements, that turn out to have strong perceptive distortions. The Hering illusion (Figure 2.1) is built on a number of converging straight lines, together with two parallel lines symmetric with respect to the convergence point. Those parallel straight lines look curved to all observers in frontal view. Although some perspective explanation (and many others) have been attempted for this illusion, it must be said that it has remained a mystery.

The same happens with the Sander and the Müller-Lyer illusions, which may also obey some perspective interpretation. In the Sander illusion, one can see an isosceles triangle *abc* (Figure 2.2(b)) inscribed in a parallelogram (Figure 2.2(a)). In Figure 2.2(a) the segment $[a, b]$ is perceived as smaller than the segment $[b, c]$. Let us attempt a perspective explanation. When we see Figure 2.2(a), we actually

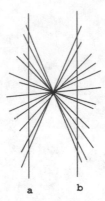

Fig. 2.1 Hering illusion: The straight lines *a* and *b* look curved in the neighborhood of a vanishing point.

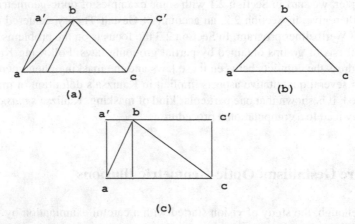

Fig. 2.2 Sander illusion: In (a), the segment [*a*,*b*] looks smaller than the segment [*b*,*c*]. Now, the isosceles triangle *abc* is the same in (a) and (b). A perspective interpretation of (a) like the one suggested in (c), where the parallelogram is thought of as a rectangle, might give some hint.

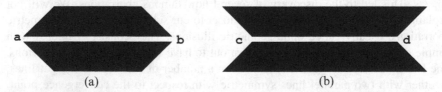

Fig. 2.3 Müller-Lyer illusion: The segment [*a*,*b*] looks smaller than [*c*,*d*].

automatically interpret the parallelogram as a rectangle in slanted view. In this interpretation, the physical length *ab* should indeed be shorter than *bc* (Figure 2.2(c)).

A hypothetical compensation mechanism, activated by a perspective interpretation, might explain the Müller-Lyer illusion as well (Figure 2.3). Here, the segments [*a*,*b*] and [*c*,*d*] have the same length but [*a*,*b*] looks shorter than [*c*,*d*]. In the per-

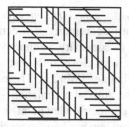

Fig. 2.4 Zoellner illusion: The diagonals inside the square are parallel but seem to alternately converge or diverge.

spective interpretation of these figures (where the trapezes are in fact rectangles in perspective), $[a, b]$ would be closer to the observer than $[c, d]$ and this might entail a difference in our appreciation of their size as actual physical objects.

As the Hering illusion, the Zoellner illusion (Figure 2.4) has parallel lines, but this time they sometimes look converging and sometimes diverging. Clearly, our global interpretation of their direction is influenced by the small and slanted straight segments crossing them. In all of these cases, one can imagine such explanations, or quite different ones based on the cortical architecture. No final explanation seems for the time being to account for all objections.

2.2 Grouping Laws and Gestalt Principles

Gestalt Theory does not continue on the same line. Instead of wondering about such or such distortion, gestaltists more radically believe that any percept is a visual illusion no matter whether or not it is in good agreement with the physical objects. The question is not why we sometimes see a distorted line when it is straight; the question is why we do see a line at all. This perceived line is the result of a construction process whose laws it is the aim of Gestalt Theory to establish.

2.2.1 Gestalt Basic Grouping Principles

Gestalt Theory starts with the assumption of active grouping laws in visual perception [Kan97, Wer23]. These groups are identifiable with subsets of the retina. We will talk in the following of points or groups of points that we identify with spatial parts of the planar rough percept. In image analysis we will identify them as well with the points of the digital image. Whenever points (or previously formed groups) have one or several characteristics in common, they get grouped and form a new, larger visual object, a *gestalt*. The list of elementary grouping laws given by Gaetano Kanizsa in *Grammatica del Vedere*, page 45ff [Kan97] is *vicinanza, somiglianza, continuita di direzione, completamento amodale, chiusura, larghezza constante, tendenza alla convessita, simmetria, movimento solidale*, and *esperienza passata* – that is, vicinity, similarity, continuity of direction, amodal completion, closure, constant width, tendency to convexity, symmetry, common motion, and past experience. This list is actually very close to the list of grouping laws considered in the founding paper by Wertheimer [Wer23]. These laws are supposed to be at work for every new percept. The amodal completion – one of the main subjects of Kanizsa's books – is, from the geometric viewpoint, a variant of the good continuation law. (The good continuation law has been extensively addressed in Computer Vision, first by Montanari in [Mon71], later by Sha'Ashua and Ullman in [SU88], and more recently by Guy and Medioni in [GM96]. An example of a Computer Vision paper implementing "good continuation", understood as being a "constant curvature", is the paper by Wuescher and Boyer [WB91]).

The *color constancy law* states that connected regions where luminance (or color) does not vary strongly are unified (seen as a whole, with no inside parts). For example, Figure 2.5 is seen as a single dark spot. The *vicinity law* applies when distance between objects is small enough with respect to the rest (Figure 2.6).

The *similarity law* leads us to group similar objects into higher-scale objects. See Figures 2.7 and 2.8. But probably one of the most pregnant and ancient constitution laws is Rubin's *closure law*, which leads us to see as an object the part of the plane

Fig. 2.5 With the color constancy law we see here a single dark spot rather than a number of dark dots.

Fig. 2.6 The vicinity law entails the grouping of the dark ellipses into two different objects.

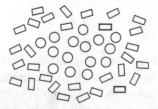 

Fig. 2.7 The similarity law leads us to interpret this image as composed of two homogeneous regions: one in the center made of circles and a peripheral one built of rectangles.

Fig. 2.8 The similarity law separates this image into two regions with different "textures". Contrarily to what happens in Figure 2.7, the shape of the group elements (squares) is not immediately apparent because of a masking effect (see Section 2.2.2).

Fig. 2.9 Because of Rubin's closure law, the interior of the black curve is seen as an object and its exterior as the background.

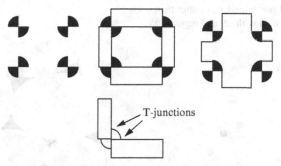

Fig. 2.10 T-junctions entail an amodal completion and a completely different image interpretation.

surrounded by a closed contour. The exterior part of the plane is then assimilated to a background. As can be appreciated in Figure 2.9, an illusory color contrast between foreground and background is often perceived.

The *amodal completion* law applies when a curve stops on another curve, thus creating a "T-junction". In such a case, our perception tends to interpret the interrupted curve as the boundary of some object undergoing occlusion. The leg of the T is then extrapolated and connected to another leg in front whenever possible. This fact is illustrated in Figure 2.10 and is called "amodal completion". The connection of two T-legs in front obeys the "good continuation" law. This means that the re-created amodal curve is as similar as possible to the pieces of curve it interpolated (same direction, curvature, etc.).

In Figure 2.10 we see first four black butterfly-like shapes. By superposing on them four rectangles, thanks to the amodal completion law, the butterflies are perceptually completed into disks. By adding instead a central white cross to the butterflies, the butterflies contribute to the perception of an amodal black rectangle. In all cases, the reconstructed amodal boundaries obey the *good continuation law*, namely they are as homogeneous as possible to the visible parts (circles in one case, straight segments in the other).

"X-junctions" may also occur and play a role as a gestalt reconstruction tool. When two regular curves cross in an image, the good continuation law leads us to see two overlapping boundaries and a *transparency phenomenon* occurs (Figure 2.11). Each boundary may be seen as the boundary of a transparent object across which the boundary of the other one still is visible. Thus, instead of dividing the image into

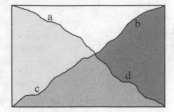

Fig. 2.11 The transparency phenomenon in the presence of an "X"-junction: We see two overlapping regions and two boundaries rather than four: region (a) is united with (d) and region (c) with (b).

Fig. 2.12 Two parallel curves: The width constancy law applies.

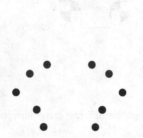

Fig. 2.13 Perceptive grouping by symmetry.

Fig. 2.14 White ovals on black background or black triangles on white background? The convexity law favors the first interpretation.

four regions, our perception only divides it into two overlapping regions bounded by both curves of the "X".

The *constant width law* applies to group the two parallel curves, perceived as the boundaries of a constant width object (Figure 2.12). This law is constantly in action since it is involved in the perception of writing and drawing.

The *symmetry law* applies to group any set of objects that is symmetric with respect to some straight line (Figure 2.13).

The *convexity law*, as the *closure law*, intervenes in our decision on the figure-background dilemma. Any convex curve (even if not closed) suggests itself as the boundary of a convex body. Figure 2.14 strikingly evidences the strength of this law and leads us to see illusory convex contours on a black background.

The *perspective law* has several forms. The simplest one was formalized by the Renaissance architect Brunelleschi. Whenever several concurring lines appear in an image, the meeting point is perceived as a vanishing point (point at infinity) in a 3-D scene. The concurring lines are then perceived as parallel lines in space (Figure 2.15).

There is no more striking proof of the strength of gestalt laws than the invention of "impossible objects". In such images, gestalt laws lead to an interpretation incompatible with physical common sense. Such is the effect of T-junctions in the famous "impossible" Penrose triangle and fork (Figures 2.16 and 2.17).

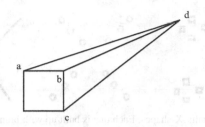

Fig. 2.15 The Y-junctions and the vanishing point *d* yield a 3-D-interpretation of this figure.

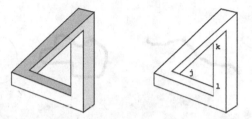

Fig. 2.16 The Penrose "impossible" triangle. Notice the T- and Y-junctions near the corners *j*, *k*, and *l*.

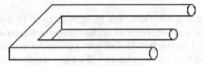

Fig. 2.17 The impossible Penrose fork. Hiding the left-hand part or the right-hand part of it leads to different perspective interpretations.

2.2.2 Collaboration of Grouping Laws

Figure 2.18 illustrates many of the grouping laws stated above. Most people would describe such a figure as "three letters X" built in different ways.

Most grouping laws stated above work from local to global. They are of mathematical nature, but must actually be split into more specific grouping laws to receive a mathematical and computational treatment:

- *Vicinity*, for instance, can mean: connectedness (i.e. spots glued together) or clusters (spots or objects that are close enough to each other and apart enough from the rest to build a group). This vicinity gestalt is at work in all subfigures of Figure 2.19.
- *Similarity* can mean: similarity of color, shape, texture, orientation, and so forth. Each one of these gestalt laws is very important by itself (see Figure 2.19).
- *Continuity of direction* can be applied to an array of objects (Figure 2.19). Let us add to it *alignments* as a grouping law by itself (constancy of direction instead of continuity of direction).
- *Constant width* is also illustrated in Figure 2.19 and is very relevant for drawings and all kinds of natural and artificial form.

Fig. 2.18 Building up a gestalt: X-shapes. Each one is built up with branches that are themselves groups of similar objects; the objects, rectangles or circles are complex gestalts, since they combine color constancy, constant width, convexity, parallelism, past experience, and so forth.

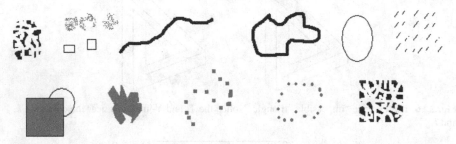

Fig. 2.19 Illustration of gestalt laws. From left to right and top to bottom: color constancy + proximity; similarity of shape and similarity of texture; good continuation; closure (of a curve); convexity; parallelism; amodal completion (a disk seen behind the square); color constancy; good continuation again (dots building a curve); closure (of a curve made of dots); modal completion – we tend to see a square in the last figure and its sides are seen in a modal way (subjective contour). Notice also the texture similarity of the first and last figures. Most of the figures involve constant width. In this complex figure, the subfigures are identified by their alignment in two rows and their size similarity.

– Notice in the same spirit that *convexity,* also illustrated, is a particularization of both closure and good continuation laws.
– *Past experience*: In the list of partial gestalts that are looked for in any image, we can have generic shapes such as circles, ellipses, rectangles, and also silhouettes of familiar objects such as faces, cats, chairs, and so forth.

All of the above listed grouping laws belong, according to Kanizsa, to the so-called *processo primario* (primary process), opposed to a more cognitive secondary process. Also, it may of course be asked *why and how* this list of geometric qualities has emerged in the course of biological evolution. Brunswick and Kamiya [BK53] were among the first to suggest that the gestalt grouping laws were directly related to the geometric statistics of the natural world. Since then, several works have addressed, from different viewpoints, these statistics and the building elements that should be conceptually considered in perception theory and/or numerically used in Computer Vision [BS96], [OF96], [GPSG01], [EG02].

The grouping laws usually collaborate to the building up of larger and larger objects. A simple object such as a square whose boundary has been drawn in black

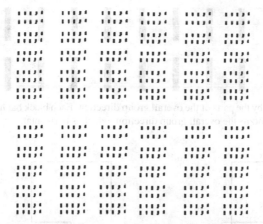

Fig. 2.20 Recursivity of gestalt laws: Here, constant width and parallelism are applied at different levels in the building up of the final group not less than six times, from the smallest bricks, which are actually complex gestalts, being roughly rectangles, up to the final rectangle. Many objects can present deeper and more complex constructions.

with a pencil on a white sheet will be perceived by connectedness (the boundary is a black line), constant width (of the stroke), convexity and closure (of the black pencil stroke), parallelism (between opposite sides), orthogonality (between adjacent sides), and again constant width (of both pairs of opposite sides).

We must therefore distinguish between *global* gestalt and *partial* gestalt. A square alone is a global gestalt, but it is the synthesis of a long list of concurring local groupings, leading to parts of the square endowed with some gestalt quality. Such parts we will call *partial gestalts*. The sides and corners of the square are therefore partial gestalts.

Notice also that all grouping gestalt laws are *recursive*: They can be applied first to atomic inputs and then in the same way to partial gestalts already constituted. Let us illustrate this by an example. In Figure 2.20 the same partial gestalt laws, namely alignment, parallelism, constant width, and proximity, are recursively applied not less than six times: the single elongated dots first aligned in rows, these rows in groups of two parallel rows, these groups again in groups of five parallel horizontal bars, these groups again in groups of six parallel vertical bars. The final groups appear to be again made of two macroscopic horizontal bars. The whole organization of such figures is seeable at once.

2.2.3 Global Gestalt Principles

Although the partial, recursive, grouping gestalt laws do not bring as much doubt about their definition as a computational task from atomic data, the global gestalt principles are by far more challenging. For many of them we do not even know

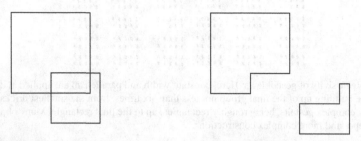

Fig. 2.21 Inheritance by the parts of the overall group direction: Each black bar has its own vertical orientation but also inherits the overall group direction, which is horizontal.

Fig. 2.22 Tendency to structural coherence and maximal regularity: The left figure is interpreted as two overlapping squares and not as the juxtaposition of the two irregular polygons on the right.

whether they are properly constitutive principles or an elegant way of summarizing various perception processes. They constitute, however, the only cues we have about the way the partial gestalt laws could be derived from a more general principle. On the other hand, these principles are absolutely necessary in the description of the perception process since they should fix the way grouping laws interact or compete to create the final global percepts – that is, the final gestalts. Let us go on with the gestalt principles list that can be extracted from [Kan97].

- *Inheritance by the parts of the overall group direction (ragruppamento secondo la direzionalita della struttura)*, Kanizsa, *Grammatica del Vedere* [Kan97] p. 54. This is a statement that might find its place in Plato's Parmenides: "the parts inherit the whole's qualities". See Figure 2.21 for an illustration of this principle.
- *Pregnancy, structural coherence, unity (pregnanza, coerenza strutturale, carattere unitario, [Kan97] p. 59), tendency to maximal regularity ([Kan97] p. 60), articulation whole/parts (in German, Gliederung), articulation without remainder ([Kan97] p. 65)*. These seven gestalt laws are not partial gestalts; in order to deal with them from the Computer Vision viewpoint, one has to assume that all partial grouping laws have been applied and that a synthesis of the groups into the final global gestalts must be thereafter performed. Each principle describes some aspect of the synthesis made from partial grouping laws into the most wholesome, coherent, complete, and well-articulated percept. See Figure 2.22 for an illustration of this principle of structural coherence.

2.3 Conflicts of Partial Gestalts and the Masking Phenomenon

With the computational discussion in mind, we wish to examine the relationship between two important technical terms of Gestalt Theory, namely *conflicts* and *masking*.

2.3.1 Conflicts

The gestalt laws are stated as independent grouping laws. They start from the same building elements. Thus, conflicts between grouping laws can occur and therefore also conflicts between different interpretations. These different interpretations may lead to the perception of different and sometimes incompatible groups in a given figure. Here are three cases.

(a) Two grouping laws act simultaneously on the same elements and give rise to two overlapping groups. It is not difficult to build figures where this occurs, as in Figure 2.23. In this example, we can group the black dots and the white dots by similarity of color. All the same, we see a rectangular grid made of all the black dots and part of the white ones. We also see a good continuing curve with a loop made of white dots. These groups do not compete.

(b) Two grouping laws compete and one of them wins. The other one is inhibited. This case is called *masking* and will be discussed thoroughly in Section 2.3.2.

(c) Conflict: In that case, both grouping laws are potentially active, but the groups cannot exist simultaneously. In addition, none of the grouping laws wins clearly. Thus, the figure is ambiguous and presents two or more possible interpretations.

A large section of Kanizsa's second chapter [Kan97] is dedicated to gestalt conflicts. Their study leads to the invention of tricky figures where an equilibrium is maintained between two conflicting gestalt laws struggling to give the final figure organization. The viewers can see both organizations and perceive their conflict. A seminal experiment due to Wertheimer [Wer23] gives an easy way to construct such

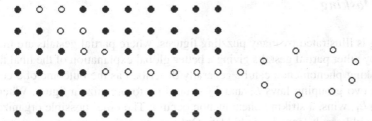

Fig. 2.23 Gestalt laws in simultaneous action without conflict: the white dots are elements of the grid (alignment, constant width) and simultaneously belong to a good continuing curve.

Fig. 2.24 Conflict of similarity of shapes with vicinity. We can easily view the left-hand figure as two groups by shape similarity: one made of rectangles and the other one of ellipses. On the right, two different groups emerge by vicinity. Vicinity "wins" against similarity of shapes.

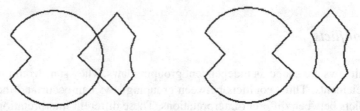

Fig. 2.25 A "conflict of gestalts": Do we see two overlapping closed curves or, as suggested on the right, two symmetric curves that touch at two points? We can interpret this experiment as a masking of the symmetry law by the good continuation law. (From Kanizsa [Kan97] p. 195.)

conflicts. In Figure 2.24 we see on the left a figure made of rectangles and ellipses. The prominent grouping laws are as follows: (a) shape similarity, which leads us to group the ellipses together and the rectangles as two conspicuous groups; (b) the vicinity law, which makes all of these elements build a unified cluster. Thus, on the left figure both laws coexist without real conflict. On the right figure, however, two clusters are present. Each one is made of heterogeneous shapes, but they fall apart enough to enforce the splitting of the ellipse group and of the rectangle group. Thus, on the right, the vicinity law dominates. Such figures can be varied by changing, for example, the distance between clusters until the final figure presents a good equilibrium between conflicting laws.

Some laws, like good continuation, are so strong that they almost systematically win, as is illustrated in Figure 2.25. Two figures with a striking axial symmetry are concatenated in such a way that their boundaries are put in "good continuation". The result is a different interpretation where the symmetric figures literally disappear. This is a conflict, and one with a total winner. It therefore is in the masking category.

2.3.2 Masking

Masking is illustrated by many puzzling figures, where partial gestalts are literally hidden by other partial gestalts giving a better global explanation of the final figure. The masking phenomenon can be generally described as the outcome of a conflict between two grouping laws L_1 and L_2 struggling to organize a figure. When one of them, L_1, wins, a striking phenomenon occurs: The other possible organization, which would result from L_2, is hidden. Only an explicit comment can remind the viewer of the existence of the possible organization under L_2: The parts of the figure that might be perceived by L_2 have become invisible, *masked* in the final figure, which is perceived under L_1 only.

Kanizsa considers four kinds of masking: *masking by embedment in a texture; masking by addition (the Gottschaldt technique); masking by subtraction (the Street technique); masking by manipulation of the figure-background articulation.* This last manipulation is central in Rubin's theory [Rub15] and in the famous Escher's drawings. The first technique we will consider is *masking in texture.* Its principle is a geometrically organized figure embedded into a texture –that is, a whole region made of similar building elements. This masking may well be called *embeddedness* as suggested by Kanizsa in [Kan91] p. 184. Figure 2.26 gives a good instance of the power of this masking, which has been thoroughly studied by the schools of Beck and Juslesz [BJ83]. In this clever figure, the basis of a triangle is literally hidden in a set of parallel lines. We can interpret the texture masking as a conflict between an arbitrary organizing law L_2 and the similarity law, L_1. The masking technique works by multiple additions embedding a figure F organized under some law L_2 into many elements that have a shape similar to the building blocks of F.

The same masking process is at work in Figure 2.27. A curve made of roughly aligned pencil strokes can be embedded and masked in a set of many more parallel strokes.

In the *masking by addition* technique due to Gottschaldt, a figure is concealed by the addition of new elements, which create another more powerful organization. Here, L_1 and L_2 can be any organizing law. In Figure 2.28, a hexagon is thoroughly concealed by the addition to the figure of two parallelograms that include in their sides the initial sides of the hexagon. Noticeably, the "winning laws" are the same that made the hexagon so conspicuous before masking, namely closure, symmetry, convexity, and good continuation.

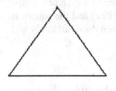 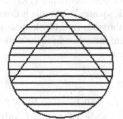 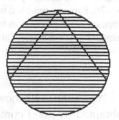

Fig. 2.26 Masking by embedding in a texture. The basis of the triangle becomes invisible as it is embedded in a group of parallel lines. (Galli and Zama, quoted in [Kan91]).

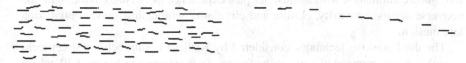

Fig. 2.27 Masking by embedding in a texture. On the right is a curve created from strokes by "good continuation". This curve is present, but masked on the left. This can be thought of as a conflict between L_2, "good continuation" and L_1, similarity of direction. The similarity of direction is more powerful because it organizes the full figure *(articulazione senza resti principle)*.

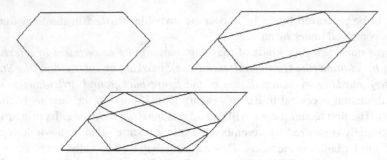

Fig. 2.28 Masking by concealment (Gottschaldt 1926). The hexagon on the left is concealed in the figure on the right and still more concealed in the bottom figure. The hexagon was built by the closure, symmetry, and convexity gestalt laws. The same laws plus the good continuation form the winner figures. They are all parallelograms.

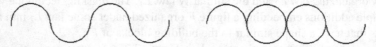

Fig. 2.29 Masking of circles in good continuation or conversely masking of good continuation by closure and convexity. We do not really see arcs of circles on the left, although significant and accurate parts of circles are present: We see a smooth curve. Conversely, we do not see the left "good" curve as a part of the right figure. It is nonetheless present.

Fig. 2.30 Masking by the Street subtraction technique (1931), inspired from Kanizsa [Kan91] p. 176. Parts are removed from the black square. When this is done in a coherent way, a new shape appears (a rough cross in the second subfigure, four black spots in the last one) and the square is masked. It is not masked at all in the third though, where the removal has been done randomly and does not yield a competing interpretation.

As Figure 2.29 shows, L_1 and L_2 can reverse their roles. On the right, the curve obtained by good continuation is made of perfect half-circles concatenated. This circular shape is masked in the good continuation. Surprisingly enough, the curve on the left is present in the figure on the right, but masked by the circles. Thus, on the left, good continuation wins against our past experience of circles. On the right, the converse occurs; convexity, closure and circularity win against good continuation and mask it.

The third masking technique considered by Kanizsa is subtraction (Street technique) – that is, removal of parts of the figure. As is apparent in Figure 2.30, where a square is amputated in three different ways, the technique is effective only when removal creates a new gestalt. The square remains in view in the third figure from the left, where the removal has been made at random and is assimilable to a random perturbation. In the second and fourth figure, the square disappears although some parts of its sides have been preserved.

We should not end this section without considering briefly the last category of masking mentioned by Kanizsa: the masking by inversion of the figure-background relationship. This kind of masking is well known thanks to the famous Escher drawings. Its principle is "the background is not a shape" (*il fondo non é forma*). Whenever strong gestalts are present in an image, the space between those conspicuous shapes is not considered as a shape, even when it has a familiar shape like a bird, a fish, or a human profile. Again here, we can interpret masking as the result of a conflict of two partial gestalt laws: one building the form and the other one, the loser, not allowed to build the background as a gestalt.

2.4 Quantitative Aspects of Gestalt Theory

In this section we open the discussion on quantitative laws for computing partial gestalts. We shall first consider some numerical aspects of Kanizsa's *masking by texture*. We shall also make some comments on Kanizsa's paradox and its answer pointing out the involvement of a quantitative image resolution. These comments lead to Shannon's sampling theory.

2.4.1 Quantitative Aspects of the Masking Phenomenon

In his fifth chapter of *Vedere e pensare* [Kan91], Kanizsa points out that "it is reasonable to imagine that a black homogeneous region contains all theoretically possible plane figures, in the same way as with Michelangelo a marble block virtually contains all possible statues." Thus, these virtual statues could be considered masked. This is after Vicario called *Kanizsa's paradox*. Figure 2.31 shows that one can obtain any simple enough shape by pruning a regular grid of black dots. In order to go further, it seems advisable to the mathematician to make a count. How many squares could we see, for example, in such a figure? Characterizing the square by its upper

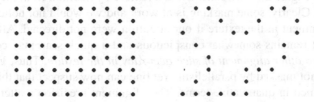

Fig. 2.31 According to Kanizsa's paradox, the figure on the right is potentially present in the figure on the left and would indeed appear if we colored the corresponding dots. This illustrates the fact that the figure on the left contains a huge number of possible different shapes.

left corner and its side length, the number of squares whose corners lie on the grid is roughly 400. The number of curves with "good continuation" made of about 20 points like the one drawn on the right of Figure 2.31 is still much larger. One indeed has 80 choices for the first point and about 5 points among the neighbors for the second point, and so forth. Thus, the number of possible good curves in our figure is grossly 80×5^{20} if we accept the curve to turn strongly and about 80×3^{20} if we ask the curve to turn at a slow rate. In both cases, the number of possible "good" curves in the grid is huge.

This multiplicity argument suggests that a grouping law can be active in an image only if its application would not create a huge number of partial gestalts. To put it another way, the multiplicity implies a masking by texture. Masking of all possible good curves in the grid of Figure 2.31 occurs just because too many such curves are possible. In Figure 2.27, we can use the same quantitative argument. In this figure the left-hand set of strokes actually invisibly contains the array of strokes on the right. This array of strokes is obviously organized as a curve (good continuation gestalt). This curve becomes invisible on the left-hand figure just because it gets endowed in a more powerful gestalt, namely parallelism (similarity of direction). As we will see in the computational discussion, the fact that the curve has been masked is related to another fact that is easy to check on the left-hand part of the figure: Many curves of the same kind as the one given on the right can be selected.

In short, we do not consider Kanizsa's paradox a hard problem but, rather, an arrow pointing toward the computational formulation of gestalt: We will define a partial gestalt as *a structure that is not masked in texture.*

We will therefore not rule out the extreme masking case, in contradiction to Vicario's principle *é mascherato solo ciò che può essere smascherato* (masked is only what can be unmasked). Clearly, all psychophysical masking experiments must be close enough to the "conflict of gestalts" situation, where the masked gestalt is still attainable when the subject's attention is directed. Thus, psychological masking experiments must remain close to the nonmasking situation and therefore satisfy Vicario's principle. Yet from the computational viewpoint, Figures 2.31 and 2.27 are nothing but very good masking examples.

In this masking issue, one feels the necessity to go from qualitative to quantitative arguments since a gestalt can be more or less masked. How to compute the right information to quantize this "more or less"? It is actually related to a *precision parameter.* In Figure 2.32 we constructed a texture by addition from the alignment drawn below. Clearly, some masking is at work and we would not notice immediately the alignment in the texture if our attention were not directed. All the same, the alignment remains somewhat conspicuous and a quick scan may convince us that *there is no other alignment of such accuracy in the texture.* Thus, in this case, alignment is not masked by parallelism. Yet one can now suspect that this situation can be explained in quantitative terms. The alignment precision matters here and should be evaluated. Precision will be one of the three parameters we shall use when computing gestalts in digital images.

Fig. 2.32 Bottom: an array of roughly aligned segments. Above the same figure is embedded into a texture in such a way that it still is visible as an alignment. We are in the limit situation associated with Vicario's proposition: "masked is only what can be unmasked".

2.4.2 Shannon Theory and the Discrete Nature of Images

The preceding subsection introduced two of the parameters we will have to deal with in computations, namely the *number of possible partial gestalts* and a *precision* parameter. Before proceeding to any computation, the computational nature of digital and biological images as raw datum must be clarified. Kanizsa addresses briefly this problem in the fifth chapter of *Vedere e pensare* [Kan91], in his discussion of the masking phenomenon: "We should not consider elements to be masked, which are too small to attain the visibility threshold." Kanizsa was aware that the number of visible points in a figure is finite: *"non sono da considerare mascherati gli elementi troppo piccoli per raggiungere la soglia della visibilita"*. He explains in the same chapter why this leads to working with figures made of dots. This decision can be seen as a way to quantize geometric information.

In order to *define* mathematically an image – be it digital or biological – in the simplest possible way, we just need to fix a point of focus. Assume all photons converging toward this focus are intercepted by a surface that has been divided into regular cells, usually squares or hexagons. Each cell counts its number of photon hits during a fixed exposure time. This count gives a gray-level image – that is, a rectangular (roughly circular in biological vision) array of gray-level values on a grid. Digital images are produced by artificial retinas or CCD's, which are rectangular grids made of square captors. In the biological case, the retina is divided into hexagonal cells with growing sizes from the fovea. Thus, in all cases, a digital or biological image contains a *finite* number of values on a grid. Shannon [Sha48] made explicit the mathematical conditions under which a continuous image can be reconstructed from this matrix of values. By Shannon's theory, one can compute the gray level at *all* points, not only at the points of the grid. Of course, when we zoom in on the interpolated image, it looks blurrier: The amount of information in a digital image is indeed *finite* and the *resolution* of the image is *bounded*. The points of the grid together with their gray-level values are called *pixels*, an abbreviation for *pic*ture *el*ements.

Fig. 2.33 When the alignment present in Figure 2.32 is made less accurate, the masking by texture becomes more efficient. The precision plays a crucial role in the computational Gestalt Theory outlined in Chapter 3.

The pixels are the computational atoms from which gestalt grouping procedures can start. Now, if the image is finite, and therefore blurry, **how can we infer sure events such as lines, circles, squares, and whatsoever gestalts from discrete data?** If the image is blurry, all of these structures cannot be inferred as completely sure; their presence and exact location must remain uncertain. This is crucial: All basic geometric information in the image has a *precision*. It is well known by gestaltists that a right angle "looks right" with some ± 3 degrees precision, and otherwise does not look right at all. Figure 2.32 shows it plainly. It is easy to imagine that if the aligned segments, still visible in the figure, are slightly less aligned, then the alignment will tend to disappear. This is easily checked with Figure 2.33, where we moved the aligned segments slightly up and down.

Let us now say briefly which local atomic information can be the starting point of computations. Since every local information about a function u at a point (x,y) boils down to its Taylor expansion, we can assume that this atomic information is:

- the value $u(x,y)$ of the gray level at each point (x,y) of the image plane. Since the function u is blurry, this value is valid at points close to (x,y);
- the gradient of u at (x,y), the vector

$$Du(x,y) = \left(\frac{\partial u}{\partial x}, \frac{\partial u}{\partial y} \right)(x,y);$$

- the *direction* at (x,y);

$$\mathrm{dir}(x) = \frac{1}{\|Du(x,y)\|} \left(-\frac{\partial u}{\partial y}, \frac{\partial u}{\partial x} \right)(x,y).$$

This vector is visually intuitive since it is tangent to the boundaries one can see in an image. This local information is known at each point of the grid and can be computed at any point of the image by Shannon interpolation. It is quantized, having a finite number of digits, and therefore noisy. Thus, each one of the preceding measurements has an intrinsic precision. The direction is invariant when the image contrast changes (which means that it is robust to illumination conditions). Bergen and Julesz [BJ83] refer to it for shape recognition and texture discrimination theory.

Gray level, gradient, and direction are the only local information we will retain for the numerical experiments in the next chapters, together with their precisions.

The aspects of Gestalt Theory that will be mathematically modeled in this book are the definition and the detection of partial gestalts such as alignment, similarity of a scalar quality (gray level, orientation, length, etc.), contrasted edges, vanishing points, and vicinity. We will see that a common framework (called a-contrario[1] detection method) can be used for all partial gestalts, but that each one of them will require a specific treatment. At the end of the book, in Chapter 14, we will present a synopsis of results with a table of formulas for all partial gestalts. It will also discuss some experiments showing how partial gestalt detectors could "collaborate". We will end this chapter with a list of unsolved questions and puzzling experiments showing the limits in the application of the found principles. In particular, the notion of "conflict" between gestalts, raised by gestaltists, has no satisfactory formal answer so far.

2.5 Bibliographic Notes

This chapter is in the form of a quick review of Gestalt Theory and contains many references. Its plan and many ideas and figures come from [DMM04] and the papers of J.-P. d'Alès, J. Froment and J.-M. Morel [dFM99], and J. Froment, S. Masnou, and J.-M. Morel [FMM98].

2.6 Exercise

2.6.1 Gestalt Essay

In this informal but important exercise, you are asked to write a gestalt commentary of the enclosed image (Figure 2.6.1). Actually any photograph can be chosen for performing the same task. The point is to forget about the meaning of the photograph and do perceptual introspection about what elementary gestalts are really perceived. It may be useful to put the photograph upside down to get a more abstract view of it.

1) Make an exhaustive list of all elementary gestalts you see in the image: curves, boundaries, alignments, parallelism, vanishing points, closed curves, convex curves, constant width objects, empirically know shapes (such as circles of letters), and symmetries.

2) Find T-junctions, X-junctions, and subjective contours.

---

[1] Because of our specific usage of the Latin expression *a contrario*, we choose to write it "a-contrario" whenever it is used as an adjective.

Fig. 2.34 A gestalt essay on this picture?

3) Discuss conflicts and masking effects between elementary gestalts as well as the figure-background dilemmas occurring in this photograph.

4) Point out the many cases where gestalts collaborate to build up objects. Discuss

Chapter 3
The Helmholtz Principle

The Helmholtz principle can be formulated two ways. The first way is common-sensical. It simply states that we do not perceive any structure in a uniform random image. In this form, the principle was first stated by Attneave [Att54]. This gestaltist was to the best of our knowledge the first scientist to publish a random noise digital image. This image was actually drawn by hand by U.S. Army privates using a random number table. In its stronger form, of which we will make great use, the Helmholtz principle states that whenever some large deviation from randomness occurs, a structure is perceived. As a commonsense statement, it states that "we immediately perceive whatever could not happen by chance". Our aim in this chapter is to discuss several intuitive and sometimes classical examples of exceptional events and their perception. We will see how hard it can be to calculate some rather simple events. This difficulty is solved by introducing a universal variable adaptable to many detection problems, the Number of False Alarms (NFA). The NFA of an event is the *expectation of the number of occurrences* of this event. Expectations are much easier to compute than probabilities because they add. After we have treated three toy examples in Section 3.1, we will define in Section 3.2 what we call *ε-meaningful events*, namely events whose NFA is less than ε. This notion is then applied to a first realistic problem: the dot alignment detection in an image.

3.1 Introducing the Helmholtz Principle: Three Elementary Examples

3.1.1 A Black Square on a White Background

Assume two scholars are looking at a picture of, say, 100×100 size, namely 10,000 pixels. Assume the figure contains somewhere a 10×10 black square; all other pixels are white (see Figure 3.1). Common sense tells us that such a figure could not arise just by chance: We are "sure" that this organization corresponds to an intention; somebody drew a square there and this is why we see it. Now, the obvious intuition

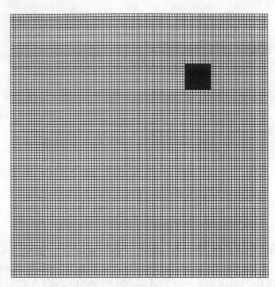

Fig. 3.1 A grid of size 100×100 containing a 10×10 square of blackened pixels; all other pixels are white. The probability of this particular configuration is $2^{-10,000}$, assuming that the pixels are independent and black or white with probability $1/2$.

that, for example, white noise cannot generate the black square must be quantified. It is actually possible that a white noise generates a black square, particularly if we are allowed to repeat the same experiment many times. As the following dialogue will show, there are many difficulties to overcome before we can "prove" the existence of the square as a meaningful event. Our dialogue takes place between a sceptic and an enthusiast. The enthusiast is sure that he sees a square and that its existence can be proven by probabilistic arguments. The sceptic will try to find, and will succeed in finding, many objections.

— Sceptic: "You think you see a square; but, all I see is a set of white or black pixels. They just fell together by chance and built this square just by chance."

— Enthusiast: "Anybody looking at a picture and knowing what a square is will claim "This is a square". Now, why are they so sure? Since you talk about chances, let us interpret their decision in a probabilistic way: Assume indeed that the pixels are white or black just by chance. Assume black and white have a probability of $1/2$. Then the probability of the black square appearing is just $(1/2)^{10,000}$ that is, about 10^{-3000}. Thus, the event is very, very, very unlikely."

— Sceptic: "Your calculation is wrong: I never said that the probabilities for white and black are equal. Any Bernoulli distribution is possible."

— Enthusiast: "Well, I am pretty sure you remember enough of your Feller reading to acknowledge that when 10,000 samples have been observed, 100 of which were black and 9900 white, then the probability of a pixel being black is likely to be close to $1/100$, is not true? So, the probability of the square happening just by chance is just about $(1/100)^{100}(99/100)^{9900}$ and it's still very small."

— Sceptic: "I concede that whatever probability p for the black you assume, the probability of the observed square is $p^{100}(1-p)^{9900}$ and therefore very small. But you've made a beginner's mistake. You know very well that probabilities must be calculated *a priori*. So you are not allowed to compute *a posteriori* the probability that particular square happened. You indeed ignored *a priori* where and what it was."

— Enthusiast: "I knew you would raise this objection! But it's easily fixed. I won't compute the probability of *that particular square happening*, but the probability of *any* square happening! Let us call k the square's side. Let's call (x,y) the position of its top-left corner: There are $(100-k+1)^2$ possible positions for this corner. Since all square events are disjoint, I can compute the probability that "some square appears somewhere in the image" and this number will be very small anyway. You will agree that the events "a square of side-length k appears at position (x,y)" are all disjoint when k varies. So I can just sum up their probabilities and I get

$$\mathbb{P}(\text{any square happening}) = \sum_{k=1}^{k=100} (100-k+1)^2 p^{k^2}(1-p)^{10,000-k^2}.$$

With $p = 1/100$, you'll agree it's a very small number anyway."

— Sceptic: "Ha! The more complicated you make it, the more objections you'll have. First of all, p depends on k. Are you forgetting that p is just an *a posteriori* estimate of the chances of a pixel being black, drawn from actual observation?"

— Enthusiast: "Well, all right then. Let's take the "unbiased estimate" of p. It's $p_k = k^2/10,000$. We can just replace p by p_k in my formula. And the sum is still very, very small."

— Sceptic: "No, it's not! Let's see... the largest term must be the first one. It's $10^4 \times 10^{-4}(1-10^{-4})^{9999} \approx e^{-1} \geq 1/3$. Do you call $1/3$ very small? Forget about your complicated formulas and use your own common sense if you have any left: If you are observing a single black pixel, are you allowed to call it a square? A dot has no shape."

— Enthusiast: "I went too far, I confess. But there must be a minimal size above which we are sure we *see* a square. So, I propose finding the *minimal size k above which we are sure we see a square*. My feeling is that if k exceeds, say 10, we are already sure we see a square. This is common sense as you call it! So I claim that the following number is very small for $k_{min} \geq 10$:

$$\mathbb{P}(\text{any square happening of size larger than } k_{min})$$
$$= \sum_{k=k_{min}}^{k=100} (100-k+1)^2 p_{k_{min}}^{k^2}(1-p_{k_{min}})^{10,000-k^2}.\text{"}$$

— Sceptic: "I'm sure you agree this computation isn't quite convincing: to start with, p is assumed to depend upon k, as $p(k) = k^2/10,000$. So you cannot fix a single k_{min}. Clearly, k_{min} depends on k. Now I'll be fair: There might be something out there. This probability you propose is indeed very small. What do you say to my next objection: The square is nothing special in your computations. You could do the same computations about *any* configuration of black pixels. So whatever random image might be presented to me, I can, by your very same computations, claim that its probability was *a priori* very low and deduce that I see something exceptional. Let me be more specific: Any realization of white noise has an equally low probability! In the case $p = 1/2$, all configurations have probability $2^{-10,000}$. So any one of them is "exceptional" along your line of thought since it has this low probability. All the same, one of them will occur. Since there are $2^{10,000}$ possible configurations, the sum of their probabilities is 1. So the event of "one of these exceptional configurations occurring" has probability $2^{10,000} \times 2^{-10,000} = 1$. No surprise there!"

— Enthusiast: "Hm, You know what? This objection poses a real problem and all the other ones were mere child's play. So I feel forced to enlarge and simplify my model. You'll agree that we do not usually recognize shapes in a white noise image. What I claim is this: The number of shapes known to humans is limited. Let us say there are as many objects as words in a good dictionary, namely 10^5. Let's assume 10^{10} aspects of the same object due to different pauses and ways it was built, angles of view, and so forth. Let's allow also for 10^3 different ways light can be shed on the same object. This means that the number of all possible black and white silhouettes of all world objects is about 10^{18}. All the same, this number is very small with respect to $2^{10,000}$. So, if we see the silhouette of a known object inside our 100×100 image, we'll immediately recognize it. The probability of each one of the familiar silhouettes occurring is $2^{-10,000}$, so the probability of any one of the silhouettes occurring is less than $2^{-10,000}.10^{18}$ which is again a very small number. So, you see, I stand my ground, since a 10×10 black square simply is one of those familiar silhouettes."

— Sceptic: "As the French say, *vous vous échappez par les branches*. We were talking about a square, and all of a sudden you start talking about *all* shapes in the world and making fantastic estimates about their number. I really don't think we're on the same page!"

3.1.2 Birthdays in a Class and the Role of Expectation

Black squares on a white background are a tough and abstract subject. So let us return to a more familiar problem: the classical problem of shared birthdays in a class. Is it surprising that two alumni have the same birthday in a class of 30? And if not, would it be surprising to observe three alumni having the same birthday? Even such a simple situation can be formalized in different ways, depending on the various answers we may put forth.

We have looked at a class of 30 students. Let us assume that their birthdays are independent and uniformly distributed variables over the 365 days of the year. We call, for $1 \leq n \leq 30$, C_n the number of n-tuples of students in the class having the same birthday (this number is computed exhaustively by considering all possible n-tuples. If, for example, students 1, 2, and 3 have the same birthday, then we count three pairs $(1,2)$, $(2,3)$, and $(3,1)$). We also consider $\mathbb{P}_n$, the probability that there is at least one n-tuple with the same birthday and p_n, the probability that there is at least one n-tuple and no $(n+1)$-tuple. In other terms, $\mathbb{P}_n$ is the probability of the event "$C_n \geq 1$", that is, "there is at least one group of n alumni having the same birthday" and p_n the probability of the event "the largest group of alumni having the same birthday has cardinality n". We are primarily interested in the evaluation of $\mathbb{P}_n$ and of the expectation $\mathbb{E}C_n$ as good indicators for the exceptionality of the event.

Proposition 1 *The probability that no two alumni have the same birthday in a class of 30 is* $(365 \times 364 \times \ldots \times 336)/(365^{30}) \approx 0.294$. *The probability that at least two alumni were born on the same day therefore is*

$$\mathbb{P}_2 \approx 0.706.$$

Proof — Number the alumni from 1 to 30. Given any date among the 365 possible, the probability that alumnus 1 has this birthday is $1/365$. So the probability that alumnus 2 has the same birthday as alumnus 1 is $1/365$. The probability that their birthdays differ therefore is $1 - 1/365 = 364/365$. In the same way the probability that alumnus 3 has a birthday different from alumni 1 and 2 is $1 - 2/365 = 363/365$. Since the birthdays are supposed independent (no twins in the class), we arrive at the expected result. □

At this point, we notice that without a computer, we would have been in some pain to compute a good approximation of this probability. There is, however, another way to demonstrate the likeliness of two alumni having the same birthday. As usual, when a probability is difficult to compute, we may compute an expectation. By the Markov inequality, expectations give hints on probabilities.

Proposition 2 *The expectation of the number of pairs of alumni having the same birthday in a class of 30 is* $\mathbb{E}C_2 = \frac{30 \times 29}{2 \times 365} \approx 1.192$. *The expectation of the number of n-tuples is* $\mathbb{E}C_n = \frac{1}{365^{n-1}} \binom{30}{n}$. *By an easy calculation,* $\mathbb{E}C_3 \approx 0.03047$ *and* $\mathbb{E}C_4 \approx 5.6 \times 10^{-4}$.

Proof — Enumerate the students from $i = 1$ to 30 and call E_{ij} the event "students i and j have the same birthday". Also, call $\chi_{ij} = \mathbb{1}_{E_{ij}}$. Clearly, $\mathbb{P}(E_{ij}) = \mathbb{E}\chi_{ij} = 1/365$. Thus, the expectation of the number of pairs of students having the same birthday is

$$\mathbb{E}C_2 = \mathbb{E}\left(\sum_{1 \leq i < j \leq 30} \chi_{ij}\right) = \sum_{1 \leq i < j \leq 30} \mathbb{E}\chi_{ij} = \frac{30 \times 29}{2} \frac{1}{365} \approx 1.192.$$

The general formula follows by analogous reasoning. □

"On the average", we can expect to see 1.192 pairs of alumni with the same birthday in each class. Unfortunately, this information is a bit inaccurate, since the large number of pairs on the average could be due to exceptional cases where one observes a lot of pairs. We only know by Markov inequality that $\mathbb{P}_2 \leq \mathbb{E}C_2$. The situation would be quite different if $\mathbb{E}C_n$ were small. In that case, an estimate on $\mathbb{E}C_n$ will give us a "cheap" estimate on $\mathbb{P}_n$. This is what we get from the estimates of $\mathbb{E}C_3$ and $\mathbb{E}C_4$. Both tell us immediately that triplets or quadruplets are not likely. They also yield an estimate of $\mathbb{P}_3$ and $\mathbb{P}_4$ from above. How good that estimate is can be derived from the following results (see Exercise 3.4.1 at the end of the chapter).

$$p_2 = \frac{1}{365^{30}} \sum_{i=1}^{15} \left[\frac{\prod_{j=0}^{i-1} \binom{30-2j}{2}}{i!} \prod_{k=0}^{29-i} (365 - k) \right] \approx 0.678$$

and, after a brave computation, $\mathbb{P}_3 \approx 0.0285$. In the same way,

$$p_3 = \frac{1}{365^{30}} \sum_{i=1}^{10} \frac{\prod_{j=0}^{i-1} \binom{30-3j}{3}}{i!}$$

$$\times \left[\prod_{k=0}^{29-2i} (365 - k) + \sum_{l=1}^{\left[\frac{30-3i}{2}\right]} \frac{\prod_{m=1}^{l} \binom{30-3i+2-2m}{2}}{l!} \prod_{n=0}^{29-2i-l} (365 - n) \right]$$

so that $p_3 \approx 0.027998$ and $\mathbb{P}_4 \approx 5.4 \times 10^{-4}$. (We denote by $[r]$ the integer part of a real number r.) To summarize, the value of $\mathbb{P}_2$ told us that it is likely to have two alumni with the same birthday. The value of $\mathbb{P}_3$ tells us that it is rare to observe triplets and the value of $\mathbb{P}_4$ tells us that quadruplets are very unlikely. It is, however, noticeable how complicated the computation of $\mathbb{P}_3$ or $\mathbb{P}_4$ has been. Of course, the formulas of $\mathbb{P}_n$ are worse and rather counterintuitive. At this point, it is noticeable how simple and intuitive the computation of $\mathbb{E}C_n$ is. For $n \geq 3$, this computation gives us exactly the same information as the computation of $\mathbb{P}_n$, namely the unlikeliness of n-uplets. More striking is that the values of $\mathbb{P}_n$ and $\mathbb{E}C_n$ differ by a very small amount. They actually give exactly the same orders of magnitude! (See the table in Exercise 3.4.1.)

3.1.3 Visible and Invisible Alignments

We return now to more visual examples. Our aim is to evaluate how well vision, put in a random environment, perceives meaningful deviations from randomness. We will try to see where the threshold stands between visible and masked alignments. On the left of Figure 3.1.3, we display roughly 400 segments. This image has size $N_1 \times N_2 = 1000 \times 600$ pixels, and the mean length of the segments is $l \simeq 30$ pixels. Thus, their directional accuracy (computed as the width-length ratio) is $\pm 2/l$, which corresponds to about ± 4 degrees. Assuming that the directions and the positions of the segments are independent and uniformly distributed, we can compute a rough estimate for the expectation of the number of alignments of four segments or more (we say that segments are aligned if they belong to the same line, up to a given

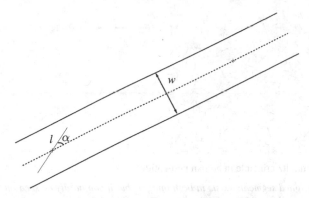

Fig. 3.2 The length l segment is said to be aligned with the dashed line with an accuracy w if the angle α is such that $\frac{l}{2}|\sin \alpha| \leq \frac{w}{2}$. Since l is much larger than w and since α is assumed to be uniformly distributed in $[-\frac{\pi}{2}, \frac{\pi}{2}]$, the probability of this event is roughly equal to $\frac{2w}{l\pi}$.

accuracy). Let M denote the number of segments, and let w denote the accuracy of the alignments (see Figure 3.2 for an illustration of this). In the following computations, we will take $w = 6$ pixels. If we consider a set of four segments denoted by S_1, S_2, S_3, and S_4, then the probability that they are aligned is roughly given by the probability that the centers of S_3 and S_4 fall at a distance less than $w/2$ from the line defined by the centers of S_1 and S_2. The relative area of the strip thus defined is approximatively $w/\max(N_1, N_2)$. Thus, this probability is $(w/\max(N_1, N_2))^2$, times the probability that the directions of the four segments are aligned with the direction of the strip. Since the segments are independent and the directions uniform, this last probability is $(2w/(l\pi))^4$. Thus, a rough estimate of the expectation of the number of alignments of four segments or more is

$$\binom{M}{4} \times \left(\frac{w}{\max(N_1, N_2)}\right)^2 \times \left(\frac{2w}{l\pi}\right)^4.$$

For the left image in Figure 3.1.3, the number of segments is $M = 400$. Using the previous formula, the expectation of the number of aligned 4-tuples of segments is about 10. It shows that we can expect some such alignments of four segments in this image. They are easily found by a computer program. Do you see them? On the right image, we performed the same experiment with about $M = 30$ segments, with the same accuracy. The expectation of the number of groups of four aligned segments is about $1/4000$. Most observers detect them immediately.

3.2 The Helmholtz Principle and ε-Meaningful Events

The three preceding examples have illustrated the promises of a general perception principle that we call the Helmholtz principle. We refer to Figure 3.2.2 for another illustration. The Helmholtz principle can be stated in the following generic way.

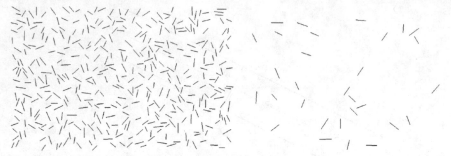

Fig. 3.3 The Helmholtz principle in human perception:

A group of four aligned segments exists in both images, but it can hardly be seen on the left-hand side image. Indeed, such a configuration is not exceptional in view of the total number of segments. In fact, the expectation of the number of aligned segments 4-tuples is about 10. In the right-hand image, we immediately perceive the alignment as a large deviation from randomness that could hardly happen by chance. In this image, the expectation of the number of groups of four aligned segments is about 1/4000.

Assume that atomic objects $O_1, O_2, \ldots, O_n$ are present in an image. Assume that k of them, say $O_1, \ldots, O_k$, have a common feature (same color, same orientation, position, etc.). We then face a dilemma: Is this common feature happening by chance or is it significant and enough to group $O_1, \ldots, O_k$? To answer this question, let us make the following mental experiment: Assume *a priori* that the considered quality had been randomly and uniformly distributed on all objects $O_1, \ldots, O_n$. In the mental experiment, the observed position of objects in the image is a random realization of this uniform process. We finally ask the question: Is the observed repartition probable or not? If not, this proves *a contrario* that a grouping process (a gestalt) is at play. The Helmholtz principle states roughly that in such mental experiments, the numerical qualities of the objects are assumed to be uniformly distributed and independent.

Definition 1 (ε-meaningful event [DMM00]). *We say that an event that is ε-meaningful if the expectation of the number of occurrences of this event is less than ε under the a-contrario random assumption. When $\varepsilon \leq 1$, we simply say that the event is meaningful.*

This definition is very generic. It must be completed by a discussion of perceptually relevant events. Adequate a-contrario models must also be given. In many cases, the a-contrario random assumption is that numerical qualities of objects are independent and uniformly distributed, but the a-contrario model can be more general.

If the Helmholtz principle is true, we perceive events if and only if they are meaningful in the sense of the preceding definition. The alignment in Figure 3.1.3 (right) is meaningful, whereas the left-hand figure contains no meaningful alignment of 4 segments.

The example of birthdays has explained why we prefer to detect unlikely events by estimating the expectation of their number instead of their probability. As an example of generic computation that we can do with the ε-meaningfulness definition, let us assume that the probability that a given object O_i has the considered quality is equal to p. In the case of birthdays, we had $p = \frac{1}{365}$ and in the black square example, $p = \frac{1}{2}$.

Under the independence assumption, the probability that at least k objects out of the observed n have this quality is

$$\mathcal{B}(n, k, p) = \sum_{i=k}^{n} \binom{n}{i} p^i (1 - p)^{n-i}$$

that is, the tail of the binomial distribution. To get an upper bound for the number of false alarms (i.e. the expectation of the number of geometric events happening by pure chance), one simply multiplies the above probability by the number of tests performed on the image. This number of tests N_{conf} corresponds to the number of different possible configurations one could have for the searched gestalt. Thus, in most cases that we will consider in the next sections, a considered event will be defined as ε-meaningful if

$$\text{NFA} = N_{conf} \cdot \mathcal{B}(n, k, p) \leq \varepsilon.$$

We call NFA the left-hand member of this inequality. It stands for "number of false alarms". The NFA of an event measures the "meaningfulness" of this event. The smaller it is, the more meaningful the event is. (Good things come in small packages.)

The definition of meaningful events is, of course, related to the statistical framework of hypothesis testing and of multiple tests. We will discuss this link and also explain the differences in Chapter 15.

3.2.1 A First Illustration: Playing Roulette with Dostoievski

Dostoievski's *The Player* is all about the links of chance and destiny. The hero of the novel believes in some regularities in chance and also believes that he can detect them and win a long series. Twice in the novel, he comments on the *exceptional* event that on some day red came in 22 times in a row, which was unheard of. We quote from [Dos69]. We translate it as follows:

That time, as if on purpose, a circumstance arose which, incidentally, recurs rather frequently in gambling. Luck sticks, for example, with red and does not leave it for ten or even fifteen turns. Only two days before, I had heard that red had come out twenty two times in a row in the previous week. One could never recall a similar case at roulette and it was spoken of with astonishment.

And earlier in the novel he writes:

In the succession of fortuitous events, there is, if not a system, at least some kind of order. (...) It's very odd. On some afternoon or morning, black alternates with red, almost without any order and all the time. Each color only appears two or three times in a row. The next day or evening, red alone turns, for example, up to twenty times in a row.

Why 22? The probability that red appears 22 times in a row is $\left(\frac{18}{37}\right)^{22}$, namely about 10^{-7}. The computation of the probability that this happens in a series of n trials may be a bit intricate. We can, instead, directly compute the expected number of occurrences of the event as $\mathrm{NFA}(n) = (n-21) \times \left(\frac{18}{37}\right)^{22}$. The event is likely to happen if its NFA is larger than 1, which yields roughly $n \geq 10^7$. Thus, we are led to compute how many trials a passionate gambler may have done in his life. Considering that a professional gambler would play roulette at 100 evenings of 5 hours a year for 20 years, estimating in addition that a roulette trial may take about 30 seconds, we deduce that an experienced gambler would observe at the most, in his gambling life span, about $n = 20 \times 100 \times 5 \times 120 \simeq 10^6$ trials. We deduce that 1 out of 10 professional gamblers can have observed such a series of 22. Actually, Dostoievski's information about the possibility of 22 series is clearly based on conversations with specialists. The hero says:

I own a good part of these observations to Mr. Astley, who spends all of his mornings by the gambling tables but never gambles himself.

If this professional observer spent his time by several tables, maybe 10 simultaneously, he is, according to our computations, likely to have observed a series of 22. As we computed, 22 is somewhat a limit for an observable series. On the other hand, the hero mentions this occurrence as having happened just a few days before he was playing. There is no contradiction here, since, according to Aristotle, it is a rule of poetry, epics, and tragedy to put their heroes in exceptional situations. As he notices in his *Poetics*, exceptional situations do happen. Dostoievski twice puts his hero in an unlikely, but not impossible, situation. First, as we mentioned, is when a series of 22 occurs just a few days before the hero gets interested in roulette, second, a few days later, is when the hero observes a series of 14 reds and takes advantage of it to win a fortune. A series of 14 is unlikely to be observed by a beginner. The NFA of this happening to the hero during the three evenings he plays at the Roulettenbourg casino is, by the same kind of calculations as above, about $\mathrm{NFA} = 3 \times 5 \times 120 \times \left(\frac{18}{37}\right)^{14} \simeq 4.10^{-3}$. Thus, this event is unlikely, yet, again, not impossible and therefore fits Aristotle's criterion.

Our comments would be incomplete if we did not also notice that the gambler's perception obeys gestalt laws. According to Dostoievski, most of their observations of roulette focus on a very small number of specific kinds of series that are clearly the only ones likely to be perceived as exceptional. These specific series are, according to Dostoievski's comments, the following:

- long monochromatic series (reds or blacks);
- periodic or quasi-periodic series, namely two or three reds alternating with two or three blacks all the time.

Thus, we can rule out the main objection raised by the sceptic of NFA calculations. He argued that all possible long sequences are equally exceptional since they all have a very low probability. There would therefore be no surprise in an exceptional one happening, since one of the sequences must happen. In fact, the observers have a very small list of gestalts and perceive all other sequences as usual and not to be noticed. Our preceding estimates should, however, take into account the number of possible gestalts, not just monochromatic series. Following Dostoievski, we can estimate to 10 the various gambler's gestalts, namely:

- long enough series of red;
- long enough series of black;
- long enough series of alternate black and red;
- long enough series of alternate pairs black-black-red-red;
- long series of alternate triples;
- long enough series alternating one red and two blacks;
- long enough series alternating one black and two reds.

There may be a few more, but little more. Let us call N_g the number of such gestalts. Then we can calculate again the NFA of the event that "any of those gestalts is observed". This NFA simply is the former NFA multiplied by N_g and our conclusions remain valid.

3.2.2 A First Application: Dot Alignments

Dots in a dot image will be called aligned if they all fall into a strip thin enough and in sufficient number (see Figure 3.2.2.) Of course, the Helmholtz a-contrario

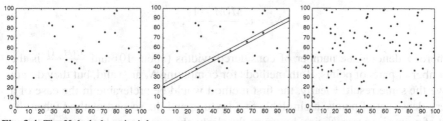

Fig. 3.4 The Helmholtz principle:

Noncasual alignments are automatically detected by the Helmholtz principle as a large deviation from randomness. Left: 20 uniformly randomly distributed dots and 7 aligned added. Middle: This meaningful and visible alignment is detected as a large deviation from randomness. Right: same alignment added to 80 random dots. The alignment is no more meaningful (and no longer visible). In order to be meaningful, it would need to contain at least 12 points.

assumption is that the dot positions are uniform, independent random variables, namely a uniform (Poisson) spatial distribution.

Let M be the number of dots in the image. The precision of the alignment is measured by the width of the strip. Let S be a strip of width a. Let $p(S)$ denote the prior probability for a point to fall in S, and let $k(S)$ denote the number of points (among the M) that are in S.

Definition 2. *A strip S is ε-meaningful if*

$$\text{NFA}(S) = N_s \cdot \mathcal{B}(M, k(S), p(S)) \leq \varepsilon,$$

where N_s is the number of considered strips.

3.2.3 The Number of Tests

We now have to discuss what the considered strips will be, since we have to evaluate their number. A simple tiling argument shows that if the strip width is small with respect to the size of the image, then $N_s \simeq 2\pi(R/a)^2$, where R is the diameter of the image domain Ω and a is the minimal width of a strip. There is indeed about that number of strips to be tested if we want to ensure that any rectangle of the image with width less than $a/2$ is contained in at least one of the strips with width a. To be more generic, we should not, however, fix an arbitrary a. So one can sample all considered strip widths a in a finite logarithmic scale up to the smallest possible width. Thus, one obtains N_s as the total number of strips of all possible (quantized) widths. Then the final number of strips N_s only depends on the size of the image and this yields an unsupervised detection method. This is the first way to compute and test the possible strips.

Second testing method. Another way to define the actual tests that speeds up detection considerably and makes it perceptually realistic is to only consider strips whose endpoints are observed dots. In such a case, we obtain

$$N_s = \alpha \frac{M(M-1)}{2},$$

where α denotes the number of considered widths (about 10) and $\frac{M(M-1)}{2}$ is the number of pairs of points. Both methods for computing N_s are valid, but they do not give the same result! Clearly, the first method would be preferable in the case of a very dense set of points, assimilable to a texture, and the second method when the set of points is sparse. Notice, however, the slight obvious change in the computation of $k(S)$. It denotes the number of dots that fell into the strip, with the exception, of course, of the two endpoints defining the strip.

At this point, we must address an objection: are we not cheating and choosing the theory that gives the better result? We have two possible values for N_s and the smallest N_s will give the largest number of detections. When two testing methods are available, perception must obviously choose the one giving the smaller test number. Indeed, there is perceptual evidence that grouping processes may depend on density and that different methods could be relevant for dense and for sparse patterns. Hence, the second testing method should be preferred for sparse distributions of points, whereas the initial model based on density would give a smaller number of tests when the number of points is large. This economy principle in the number of tests has been developed in recent works by Donald Geman and his collaborators [FG01][BG05].

Let us compare both definitions of object alignments in the examples of Figure 3.2.2. When we use the larger N_s corresponding to the all strips with all widths (from 2 to 12 pixels), *we simply do not detect any alignment*. Indeed, for this image (size 100×100), we have $R = 100\sqrt{2}$ and thus $N_s = \sum_{a=2}^{12} 2\pi (R/a)^2 \simeq 10^5$. On the other hand the alignment of 7 points is included in a strip with width $a = 3$ and thus has a probability $\mathcal{B}(M, 7, a/100)$, which has value $\simeq 10^{-5}$ when $M = 27$ and has value $\simeq 10^{-2}$ when $M = 87$. Thus, in both cases, the alignment is not meaningful. This is due to the testing overdose: We have tested many times the same alignments and have also tested many strips that contained no dots at all. The second definition of N_s happens to give a perceptually correct result. One has $N_s \simeq 3 \times 10^3$ for the image with $M = 27$ points and thus the alignment becomes meaningful. For the image with $M = 87$ points $N_s \simeq 4 \times 10^5$ and the alignment is not meaningful since its NFA is larger than 1. This result is displayed in Figure 3.2.2 in the middle, where we see the only detected strip. This same alignment is no more detectable on the right. The tested widths range from 2 to 12; strips thinner than 2 pixels are nonrealistic in natural (nonsynthetic) images and strips larger than 12 no longer give the appearance of alignments in a 100×100 image.

3.3 Bibliographic Notes

The program stated here has been proposed several times in Computer Vision. We know of at least two instances: David Lowe [Low85] and Witkin-Tenenbaum [WT83]. Here we quote extensively David Lowe's program, whose mathematical consequences are developed in this book.

We need to determine the probability that each relation in the image could have arisen by accident, $P(a)$. Naturally, the smaller that this value is, the more likely the relation is to have a causal interpretation. If we had completely accurate image measurements, the probability of accidental occurrence could become vanishingly small. For example, the probability of two image lines being exactly parallel by accident of viewpoint and position is zero. However, in real images there are many factors contributing to limit the accuracy of measurements. Even more important is the fact that we do not want to limit ourselves to perfect instances of each relation in the scene – we want to be able to use the information available from even approximate instances of a relation. Given an image relation that holds within some degree *of accuracy, we wish to calculate the probability that it could have arisen by accident to within that level of accuracy. This can only be done in the context of some assumption regarding the surrounding distribution of objects, which serves as the null hypothesis against which we judge significance. One of the most general and obvious assumptions we can make is to assume that a background of independently positioned objects in three-space, which in turn implies independently positioned projections of the objects in the image. This null hypothesis has much to recommend it. (...) Given the assumption of independence in three-space position and orientation, it is easy to calculate the probability that a relation would have arisen to within a given degree of accuracy by accident. For example if two straight lines are parallel to within 5 degrees, we can calculate that the chance is only $5/180 = 1/36$ that the relation would have arisen by accident from two independent objects.*

Some main points of the program that we will mathematically develop are contained in the preceding quotation, particularly, the idea that significant geometric objects are the ones with small probability and the idea that this probability is anyway never zero because of the inherent lack of accuracy of a digital image. However, the preceding program is not accurate enough to give the right principles for computing gestalt. The above-quoted example is not complete. Indeed we simply cannot fix *a priori* an event such as "these two lines are parallel" without merging it into the set of all events of the same kind – that is, all possible groups of parallel lines in the considered image. If the image has many lines, it simply likely that two of them will be quite parallel. So we have to take into account the number of possible pairs of parallel lines. If this number is large, then we will, in fact, detect many nonsignificant pairs of parallel lines. Only if the expected number of such pairs is much below 1, can one decide that the observed parallelism makes sense. Although, in accordance with the former quotation, the general principle proposed in this chapter should be attributed to Lowe, it is also stated by Zhu in [Zhu99] and attributed to Helmholtz [vH99]: *Besides Gestalt Psychology, there are two other theories for perceptual organization. One is the likelihood principle [vH99] which assigns a high probability for grouping two elements such as line segments, if the placement of the two elements has a low likelihood of resulting from* accidental arrangement. Viewed that way, the Helmholtz principle is exactly opposite to the so-called *Prägnanz* principle in gestalt psychology : *"...of several geometrically possible organizations that one will actually occur which possesses the best, simplest and most stable shape"*, quoted in [Zhu99] from Koffka's book [Kof35].

3.4 Exercise

3.4.1 Birthdays in a Class

Consider a class of 30 students and assume that their birthdays are independent and uniformly distributed variables over the 365 days of the year. We call, for $1 \leq n \leq 30$, C_n the number of n-tuples of students of the class having the same birthday.

(This number is computed exhaustively by considering all possible n-tuples. If (for example) students 1, 2, and 3 have the same birthday, then we count three pairs, $(1,2)$, $(2,3)$, $(3,1)$.) We also consider $\mathbb{P}_n = \mathbb{P}(C_n \geq 1)$, the probability that there is at least one n-tuple with the same birthday and p_n, the probability that there is at least one n-tuple and no $(n+1)$-tuple.

1) Prove that $\mathbb{P}_n = 1 - \sum_{i=1}^{n-1} p_i$ and $\mathbb{P}_n = \mathbb{P}_{n-1} - p_{n-1}$.

2) Prove that $\mathbb{E}C_n = \frac{1}{365^{n-1}} \binom{30}{n}$. Check that $\mathbb{E}C_2 \approx 1.192$, $\mathbb{E}C_3 \approx 0.03047$, and $\mathbb{E}C_4 \approx 5.6 \times 10^{-4}$.

3) Prove that $\mathbb{P}(C_2 = 0) = \frac{365 \times \cdots \times 336}{365^{30}} \approx 0.294$. Deduce that $P_2 \approx 0.706$.

4) Prove that

$$p_2 = \frac{1}{365^{30}} \sum_{i=1}^{15} \frac{\prod_{j=1}^{i} \binom{32-2j}{2}}{i!} \prod_{k=0}^{29-i} (365 - k).$$

5) Compute by a small computer program (in Matlab for example): $p_2 \approx 0.678$.

6) Deduce that $\mathbb{P}_3 \approx 0.0285$.

7) We denote by $[r]$ the integer part of a real number. Prove that

$$p_3 = \frac{1}{365^{30}} \sum_{i=1}^{10} \frac{\prod_{j=1}^{i} \binom{33-3j}{3}}{i!}$$

$$\times \left[\prod_{k=0}^{29-2i} (365 - k) + \sum_{l=1}^{\left[\frac{30-3i}{2}\right]} \frac{\prod_{m=1}^{l} \binom{30-3i+2-2m}{2}}{l!} \prod_{n=0}^{29-2i-l} (365 - n) \right].$$

8) Deduce by a computer program that $p_3 \approx 0.027998$ and $\mathbb{P}_4 \approx 5.4 \times 10^{-4}$.

9) Be courageous and give a general formula for p_n.

10) Prove that $\mathbb{E}C_{30} = \mathbb{P}_{30} = \frac{1}{365^{29}}$, $\mathbb{E}C_{29} = \frac{30}{365^{28}}$, and $\mathbb{P}_{29} = \frac{30 \times 364 + 1}{365^{29}}$.

11) The following table summarizes the comparative results for $\mathbb{E}C_n$ and P_n as well as the relative differences. Check it.

n	$\mathbb{E}C_n$	$\mathbb{P}_n$	$\frac{\mathbb{E}C_n - \mathbb{P}_n}{\mathbb{P}_n}$
2	1.192	0.706	68.84%
3	0.0347	0.0285	21.75%
4	5.6×10^{-4}	5.3×10^{-4}	5.66%
...	...	...	...
29	$\frac{30}{365^{28}}$	$\frac{30 \times 364 + 1}{365^{29}}$	0.27%
30	$\frac{1}{365^{29}}$	$\frac{1}{365^{29}}$	0%

12) Explain why $\mathbb{P}_n$ and $\mathbb{E}C_n$ are so close for $n \geq 3$.

Chapter 4
Estimating the Binomial Tail

We saw in the previous chapter that the Helmholtz principle in his generic form leads us to the computation of the probability of events of the type "at least k objects out of l have a considered quality". When the a-contrario assumption is that objects are independent and have the same probability p to have the quality, the probability of this event is given by the binomial distribution. In this chapter, we will give different inequalities and asymptotic results for the binomial tail. Such results are useful mainly because they help us to understand the "meaningfulness" of an event as a function of l, k, and p. The results given in this chapter will be used through the rest of the book.

4.1 Estimates of the Binomial Tail

The Helmholtz principle leads us to evaluate a number of false alarms

$$\text{NFA}(l,k,p) = N_{test} \cdot \mathbb{P}[S_l \geq k],$$

where

$$S_l = \sum_{i=1}^{l} X_i$$

and X_i are independent Bernoulli random variables with parameter p, namely $\mathbb{P}[X_i = 1] = p$ and $\mathbb{P}[X_i = 0] = 1 - p$ (see Exercise 4.3.1 at the end of the chapter). The number of tests N_{test} being generally easy to calculate (but not always, it depends on the considered geometric configurations. We will see that it requires a different computation for each different elementary gestalt), the estimation of the NFA boils down to an estimate of the tail of the binomial law,

$$\mathcal{B}(l,k,p) = \sum_{i=k}^{l} \binom{l}{i} p^i (1-p)^{l-i}.$$

In this estimation problem, p is fixed, l is rather large, and k in excess with respect to its expected value pl since we look for meaningful events. The number of tests usually will be very large and the NFA is interesting mainly when it is smaller than 1; so we are primarily interested in good estimates of $\mathcal{B}(l,k,p)$ when this quantity is very small. There are several tools available to do so.

- The first one is, of course, the **law of large numbers**, which tells us that $\mathcal{B}(l,k,p) = \mathbb{P}[S_l \geq k] \to 0$ whenever $\frac{k}{l} \geq r > p$ and $l \to \infty$. This result yields no order of magnitude for $\mathcal{B}(l,k,p)$.
- **Large deviations techniques** give asymptotic estimates of $\mathcal{B}(l,k,p)$ when k is roughly proportional to l, namely $k \simeq rl$ with $r > p$.
- The **Central Limit Theorem** also gives estimates for smaller deviations from the mean, namely when $k \simeq pl + C\sqrt{l}$.

We will state these estimates along with some more precise variants. Notice that our estimation problem is not an asymptotic problem. We need a nonasymptotic precise evaluation of $\mathcal{B}(l,k,p)$ valid if possible for all values of l,k, and p and accurate whenever $\mathcal{B}(l,k,p)$ happens to be small.

For practical purposes, it seems that we could give up any estimate of the binomial tail and simply compute exact numerical values of $\mathcal{B}(l,k,p)$. These values can even be tabulated once and for all. In practice as well as in theory we also are interested in explicit formulas of the order of magnitude of the minimal value $k(l)$ for which a gestalt with parameters $(l,k(l),p)$ becomes detectable. Such formulas will be useful, for example, in Chapter 5, which is dedicated to the detection of alignments in a digital image. They will indeed yield an accurate estimate of the minimal number $k(l)$ of aligned points in a segment of length l leading to an alignment perception. This estimate will be a simple explicit function of l and p. In Chapter 7, dedicated to the detection of modes in a histogram, l is so large that the actual computation of $\mathcal{B}$ with computers becomes cumbersome and the large deviation estimate becomes useful. Finally, in Chapters 6 and 7, we will deal with *structural* properties of meaningful gestalts which can only be proved with the close formula estimates and formal properties of $\mathcal{B}(l,k,p)$.

As we mentioned earlier, there are several theories giving asymptotic estimates of $\mathcal{B}(l,k,p)$ and, luckily enough, there also are inequalities that give good lower and upper estimates of $\mathcal{B}(l,k,p)$, namely the Hoeffding and Slud inequalities. We will mainly use the asymptotic estimates to evaluate how good these nonasymptotic inequalities are.

A unified format is needed to compare these inequalities. In what follows we will compare the estimated values of

$$-\frac{1}{l}\log\mathcal{B}(l,k,p).$$

4.1.1 Inequalities for $\mathcal{B}(l,k,p)$

Proposition 3 (Hoeffding inequalities [Hoe63]) *Let* $X_1,\ldots,X_l$ *be independent random variables such that* $0 \leq X_i \leq 1$. *We set* $S_l = \sum_{i=1}^{l} X_i$ *and* $p = \mathbb{E}\left[\frac{S_l}{l}\right]$. *Then, for* $pl < k < l$,

$$\mathbb{P}\left[S_l \geq k\right] \leq \left(\frac{p}{k/l}\right)^{l\left(\frac{k}{l}\right)} \left(\frac{1-p}{1-\frac{k}{l}}\right)^{l(1-k/l)}.$$

In other terms, setting $r = k/l$,

$$\mathbb{P}\left[S_l \geq k\right] \leq e^{-l\left(r\log\frac{r}{p}+(1-r)\log\frac{1-r}{1-p}\right)}.$$

In addition, the right-hand term of this inequality satifies

$$e^{-l\left[r\log\frac{r}{p}+(1-r)\log\frac{1-r}{1-p}\right]} \leq e^{-l(r-p)^2 h(p)} \leq e^{-2l(r-p)^2},$$

where

$$h(p) = \begin{cases} \dfrac{1}{1-2p}\log\dfrac{1-p}{p} & \text{if} \quad 0 < p < \frac{1}{2}, \\[2mm] \dfrac{1}{2p(1-p)} & \text{if} \quad \frac{1}{2} \leq p < 1. \end{cases}$$

We refer to Figure 4.1 for the graph of the function $p \mapsto h(p)$. A more general version of this inequality is proved in Exercise 4.3.2 at the end of this chapter. In the particular case where the X_i are Bernoulli variables, this inequality is also known as the Chernoff inequality [Che52].

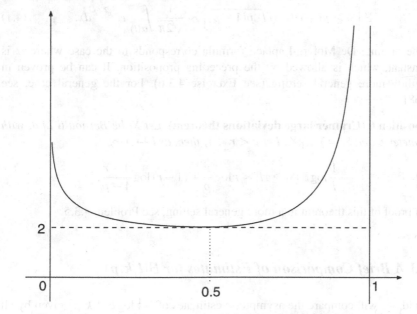

Fig. 4.1 The graph of the function $p \mapsto h(p)$ involved in Hoeffding inequality.

Theorem 1 ((Slud's Theorem [Slu77])) *If* $0 < p \le 1/4$ *and* $pl \le k \le l$, *then*

$$\mathbb{P}[S_l \ge k] \ge \frac{1}{\sqrt{2\pi}} \int_{\alpha(l,k)}^{+\infty} e^{-x^2/2} \, dx \qquad \text{where } \alpha(l,k) = \frac{k-pl}{\sqrt{lp(1-p)}}.$$

The assumption $0 < p \le 1/4$ is not a strong condition for us. It will be true for roughly all applications we consider.

4.1.2 Asymptotic Theorems for $\mathcal{B}(l,k,p) = \mathbb{P}[S_l \ge k]$

Proposition 4 (Law of large numbers) *Let* X_i *be Bernoulli i.i.d. random variables with parameter* p *and* $S_l = \sum_{i=1}^{l} X_i$. *Then*

$$\mathbb{P}\left[\frac{S_l}{l} \xrightarrow[l \to \infty]{} p \right] = 1.$$

As a consequence, for every $r > p$ *and* $l \to \infty$,

$$\mathbb{P}[S_l \ge rl] \to 0, \quad \text{that is,} \quad \mathcal{B}(l,[rl],p) \to 0$$

(where $[s]$ *denotes the integer part of a real number s.)*

Proposition 5 (Generalized Central Limit Theorem) *Let* X_i *be Bernoulli i.i.d. random variables with parameter* p *and let* $S_l = \sum_{i=1}^{l} X_i$. *If* $\alpha(l)^6/l \to 0$ *as* $l \to +\infty$, *then*

$$\mathbb{P}\left[S_l \ge pl + \alpha(l)\sqrt{l \cdot p(1-p)} \right] \sim \frac{1}{\sqrt{2\pi}} \int_{\alpha(l)}^{+\infty} e^{-x^2/2} \, dx. \qquad (4.1)$$

The original De Moivre-Laplace formula corresponds to the case where α is a constant, which is allowed by the preceding proposition. It can be proven in a slightly more general setting (see Exercise 4.3.6). For the general case, see [Fel68].

Proposition 6 (Cramér large deviations theorem) *Let* X_i *be Bernoulli i.i.d. with parameter* p *and* $S_l = \sum_{i=1}^{l} X_i$. *Let* $p < r < 1$, *then, as* $l \to +\infty$,

$$-\frac{1}{l} \log \mathbb{P}[S_l \ge rl] \sim r \log \frac{r}{p} + (1-r) \log \frac{1-r}{1-p}.$$

For a proof of this theorem in a more general setting, see Problem 4.3.5.

4.1.3 A Brief Comparison of Estimates for $\mathcal{B}(l,k,p)$

As said, we will compare the asymptotic estimates of $-\frac{1}{l} \log \mathcal{B}(l,k,p)$ given by all of the preceding propositions put in a uniform format.

Proposition 7 *Let X_i, $i = 1, \ldots, l$ be independent Bernoulli random variables with parameter $0 < p < \frac{1}{4}$ and let $S_l = \sum_{i=1}^{l} X_i$. Consider a constant $p < r < 1$ or a real function $p < r(l) < 1$. Then $\mathcal{B}(l, k, p) = \mathbb{P}[S_l \geq k]$ satisfies*

$$(Slud) \quad -\tfrac{1}{l}\log \mathbb{P}[S_l \geq rl] \ \leq \ \frac{(r-p)^2}{p(1-p)} + O(\frac{\log l}{l}), \tag{4.2}$$

$$(Hoeffding\text{-}bis) \quad -\tfrac{1}{l}\log \mathbb{P}[S_l \geq rl] \ \geq \ (r-p)^2 \frac{\log \frac{1-p}{p}}{1-2p} + O(\tfrac{1}{l}), \tag{4.3}$$

$$(Central\ limit) \ -\tfrac{1}{l}\log \mathbb{P}[S_l \geq r(l)l] \ \sim \ \frac{(r(l)-p)^2}{p(1-p)} \ \text{if}\ (r(l)-p)l^{\frac{1}{3}} \xrightarrow[l\to\infty]{} 0, \tag{4.4}$$

$$(Hoeffding) \quad -\tfrac{1}{l}\log \mathbb{P}[S_l \geq rl] \ \geq \ r\log\frac{r}{p} + (1-p)\log\frac{1-r}{1-p}, \tag{4.5}$$

$$(Large\ deviation) \ -\tfrac{1}{l}\log \mathbb{P}[S_l \geq rl] \ \sim \ r\log\frac{r}{p} + (1-p)\log\frac{1-r}{1-p}, \tag{4.6}$$

where the last equivalence holds when r is fixed and l tends to infinity.

Proof — The first announced inequality immediately follows from Slud's Theorem 1 and the following inequality:

$$\frac{x}{1+x^2}e^{-\frac{x^2}{2}} \leq \int_x^{+\infty} e^{-\frac{y^2}{2}}\,dy. \tag{4.7}$$

This inequality is proven in Exercise 4.3.7 and we have set in Slud's theorem

$$\alpha(k,l) = \frac{k-pl}{\sqrt{lp(1-p)}} = \frac{(r-p)\sqrt{l}}{\sqrt{p(1-p)}}.$$

The second and fourth inequalities immediately derive from the first and second forms of Hoeffding's inequality (Proposition 3) and the third and last inequalities are a mere reformulation of Propositions 5 and 6. More precisely, we have set in Camér's theorem, $r(l)l = pl + \alpha(l)\sqrt{lp(1-p)}$. □

We have put all asymptotic and nonasymptotic estimates in a homogeneous form that makes their comparison easier. The Slud and Hoeffding inequalities are precise nonasymptotic ones. In order to extract their main terms, we had to introduce an asymptotic term $O(\frac{1}{l})$ or $O(\frac{\log l}{l})$. Now, both terms are easily estimated with sharp constants, which prove them to be negligible when, say, l exceeds 100 independently of r and p. So in these inequalities, r and l can vary in anyway without any constraint provided that l is realistically large. In the Central Limit Theorem instead, $r(l)$ must be so that $r(l) - p$ tends to zero fast enough, but we can anyway compare $(Slud)$ and $(Central\ limit)$ under this condition: then we see that the estimates obtained are exactly the same up to some $O(\frac{\log l}{l})$. We conclude that Slud's theorem cannot be improved. The three first estimates given in Proposition 7 are of the same form and give as the main term an $(r-p)^2$ term multiplied by a p dependent function. So, as we will see in the next chapter (Chapter 5), they will yield the same orders of magnitude for the detection thresholds of alignments. Finally, the most striking fact comes from the classical comparison of $(Hoeffding)$ and $(Large\ deviation)$. Both give exactly the same estimate of the binomial tail when r is fixed and l tends to infinity. This makes a strong argument in favor of the use of the Hoeffding inequality as a general estimate for the binomial tail. We will see in Chapter 7 that this estimate has outstanding structural properties.

In the exercises at the end of the chapter, we give the proofs of two Hoeffding inequalities, of the large deviations theorem, and of the Central Limit Theorem. Such results are very general (the binomial tail is just a particular case), and thus they can be used in applications where a distribution different from the binomial one is involved.

4.2 Bibliographic Notes

Upper bounds for $\mathbb{P}[S_l \geq k]$ were obtained by Bernstein in 1946 and Bennett in 1962 [Ben62] before Hoeffding [Hoe63] proved Proposition 3 in 1963. In the case of Bernouilli i.i.d. random variables (i.e., S_l follows a binomial law), Hoeffding's inequality was implicitly contained in a paper published by Chernoff [Che52] in 1952. The second part of Hoeffding's inequality (obtained thanks to the fact that $h(p) \geq 2$ for all p) was proven by Okamoto [Oka58] in 1958 in the binomial case when $p \geq 1/2$ (which is not really the case in which we are interested).

Hoeffding seemed to close the subject since his upperbound was in a certain sense optimal. However, in 1995, Talagrand showed in [Tal95] that the exponential term of Hoeffding's inequality could be refined by means of a "missing factor."

The problem of asymptotic estimates of $\mathbb{P}[S_l \geq k]$ as l tends to infinity was addressed by Laplace (for p fixed and k in a range of the form $[lp + \alpha\sqrt{lp(1-p)}, lp + \beta\sqrt{lp(1-p)}]$). Prohorov (1961) investigates in [Pro61] this problem for all values of k and p, but in a way depending on the relative values of p and n. In 1969, Littewood [Lit69] gave an asymptotic estimate valid for p fixed and k in the range $[p, \frac{3(1-p)}{4p}]$. A review of these results and more recent and precise estimates, valid for a larger range of values for k, can be found in [Moi01], where the asymptotic results of Bahadur [Bah60] are improved for "very large deviations" ($k \simeq n$).

4.3 Exercises

4.3.1 The Binomial Law

Let $X_1, \ldots, X_n$ be i.i.d. Bernoulli variables, that is, independent variables such that $X_j \in \{0, 1\}$ and
$$\mathbb{P}[X_j = 1] = p, \quad \mathbb{P}[X_j = 0] = 1 - p = q.$$
We set $S_n = X_1 + \cdots + X_n$.

1) Prove that for $1 \leq k \leq n$,

$$\binom{n+1}{k} = \binom{n}{k} + \binom{n}{k-1}.$$

Give an interpretation of this formula.

2) Prove that for $0 \leq k \leq n$,

$$\mathbb{P}[S_n = k] = \binom{n}{k} p^k (1-p)^{n-k} = \frac{n!}{k!(n-k)!} p^k (1-p)^{n-k}.$$

This law is called the binomial law with size n and parameter p and we denote by $\mathcal{B}(n,p)$ the probability distribution defined by

$$p_k = \frac{n!}{k!(n-k)!} p^k (1-p)^{n-k}.$$

3) Compute the mean and the variance of a random variable X with binomial $\mathcal{B}(n,p)$ law (answers: np and npq.)

4.3.2 Hoeffding's Inequality for a Sum of Random Variables

Let $(\Omega, \mathcal{F}, \mathbb{P})$ be a probability space. It is recalled that if X_1 and X_2 are independent variables, and if f and g are two measurable functions from $\mathbb{R}$ to $\mathbb{R}$, then $f(X_1)$ and $g(X_2)$ are also independent. It is also recalled that if $X_1, \ldots, X_l$ are independent, then $\mathbb{E}(X_1 \ldots X_l) = (\mathbb{E}X_1) \ldots (\mathbb{E}X_l)$. Let $X_1, \ldots, X_l$ be independent random variables such that $0 \leq X_i \leq 1$. We set $S_l = \sum_{i=1}^{l} X_i$ and $p = \mathbb{E}\left[\frac{S_l}{l}\right]$. We will prove that for $0 < t < 1 - p$,

$$\mathbb{P}[S_l \geq (p+t)l] \leq \left(\frac{p}{p+t}\right)^{l(p+t)} \left(\frac{1-p}{1-p-t}\right)^{l(1-p-t)} \leq e^{-lt^2 h(p)} \leq e^{-2lt^2}, \quad (4.8)$$

where

$$h(p) = \frac{1}{1-2p} \log \frac{1-p}{p} \quad \text{if} \quad 0 < p < \frac{1}{2}$$

$$\text{and} \quad h(p) = \frac{1}{2p(1-p)} \quad \text{if} \quad \frac{1}{2} \leq p < 1.$$

This kind of inequality is called a "large deviation inequality". To understand its meaning in a more particular setting, assume that the X_i are i.i.d.. Then when l is large, we know by the law of large numbers that $\frac{S_l}{l} \to p$ almost surely. Thus, the Hoeffding inequality estimates an event that is less and less likely when l tends to infinity, namely the event that the mean $\frac{S_l}{l}$ exceeds p by a positive value t.

1) To understand the meaning of the inequality, draw the graph of the function $h(p)$.

2) Let X be a random variable such that $a \le X \le b$. Let λ be a positive real number. By using the convexity of the exponential function $x \mapsto e^{\lambda x}$, prove that

$$\mathbb{E}\left[e^{\lambda X}\right] \le \frac{b - \mathbb{E}[X]}{b - a}e^{\lambda a} + \frac{\mathbb{E}[X] - a}{b - a}e^{\lambda b}.$$

3) The main trick of large deviation estimates is to use the very simple inequality $\mathbb{1}_{x \ge 0} \le e^{\lambda x}$, true for $\lambda > 0$. Prove this inequality. Then apply it to $\mathbb{1}_{\{S_l - \mathbb{E}[S_l] - lt \ge 0\}}$ and deduce that

$$\mathbb{P}[S_l \ge (p + t)l] \le e^{-\lambda(p+t)l}\prod_{i=1}^{l}\mathbb{E}\left[e^{\lambda X_i}\right].$$

4) Set $p_i = \mathbb{E}[X_i]$. Applying question 2 with $a = 0$ and $b = 1$, deduce that

$$\prod_{i=1}^{l}\mathbb{E}\left[e^{\lambda X_i}\right] \le \prod_{i=1}^{l}(1 - p_i + p_i e^{\lambda}).$$

Be sure to check that this inequality becomes an identity when the X_i's are Bernoulli random variables.

5) Prove the geometric-arithmetic mean inequality: if $a_1, \ldots, a_l$ are positive real numbers, then

$$\left(\prod_{i=1}^{l}a_i\right)^{1/l} \le \frac{1}{l}\sum_{i=1}^{l}a_i.$$

6) Deduce that $\prod_{i=1}^{l}\mathbb{E}\left[e^{\lambda X_i}\right] \le (1 - p + pe^{\lambda})^l$.

7) Combine questions 3 and 6 and get an inequality. Prove that the right-hand side of this inequality is minimal for $\lambda = \log\frac{(1-p)(p+t)}{(1-p-t)p}$. Check that this number is positive when $0 < t < 1 - p$ and obtain the first Hoeffding inequality (left inequality in (4.8)) by taking this value for λ in the inequality.

8) To prove the second inequality of (4.8), one can remark that the first proved upper bound has a form $e^{-lt^2 G(t,p)}$, where $G(t,p)$ is defined by

$$G(t,p) = \frac{p+t}{t^2}\log\frac{p+t}{p} + \frac{1-p-t}{t^2}\log\frac{1-p-t}{1-p}.$$

Thus it is enough to show that $h(p) \le \inf_{0 < t < 1-p}G(t,p)$. First check that

$$t^2\frac{\partial}{\partial t}G(t,p) = \left(1 - 2\frac{1-p}{t}\right)\log\left(1 - \frac{t}{1-p}\right) - \left(1 - 2\frac{p+t}{t}\right)\log\left(1 - \frac{t}{t+p}\right)$$

$$= H\left(\frac{t}{1-p}\right) - H\left(\frac{t}{t+p}\right),$$

where $H(x) = (1 - \frac{2}{x})\log(1 - x)$. Continuing, show that

$$H(x) = 2 + \left(\frac{2}{3} - \frac{1}{2}\right)x^2 + \left(\frac{2}{4} - \frac{1}{3}\right)x^3 + \left(\frac{2}{5} - \frac{1}{4}\right)x^4 + \cdots$$

Deduce that $H(x)$ is increasing for $0 < x < 1$ and that $\frac{\partial}{\partial t}G(t,p) > 0$ if and only if $\frac{t}{1-p} > \frac{t}{p+t}$, which means $t > 1 - 2p$. Deduce that, when $1 - 2p > 0$, the function $G(t,p)$ attains its minimum for $t = 1 - 2p$ and the corresponding minimal value is $h(p) = \frac{1}{1-2p} \log \frac{1-p}{p}$.

9) Prove that if $1 - 2p \leq 0$, then $G(t,p)$ attains its minimum when $t \to 0$ and that in such a case,

$$\lim_{t \to 0} G(t,p) = \frac{1}{2p(1-p)} = h(p).$$

Check that $h(p) \geq h(\frac{1}{2}) = 2$ and the proof of the announced Hoeffding inequalities is complete.

4.3.3 A Second Hoeffding Inequality

We now consider independent bounded random variables with different bounds, namely $a_i \leq X_i \leq b_i$. We set $S_l = \sum_{i=1}^{l} X_i$ and $p = \mathbb{E}\left[\frac{S_l}{l}\right]$. Then we will prove that for every $t > 0$,

$$\mathbb{P}[S_l \geq (p+t)l] \leq \exp\left(-\frac{2l^2 t^2}{\sum_{i=1}^{l}(b_i - a_i)^2}\right).$$

1) Set $p_i = \mathbb{E}[X_i]$. Prove that for all $\lambda > 0$,

$$\mathbb{P}[S_l \geq (p+t)l] \leq e^{-\lambda l t} \prod_{i=1}^{l} \mathbb{E}\left[e^{\lambda(X_i - p_i)}\right].$$

2) Using question 2 of Exercise 4.3.2, show that

$$\mathbb{E}\left[e^{\lambda(X_i - p_i)}\right] \leq e^{-\lambda(p_i - a_i)}\left(\frac{b_i - p_i}{b_i - a_i} + \frac{p_i - a_i}{b_i - a_i}e^{\lambda(b_i - a_i)}\right).$$

3) Set $\lambda_i = \lambda(b_i - a_i)$ and $q_i = \frac{p_i - a_i}{b_i - a_i}$. We denote by $L(\lambda_i)$ the logarithm of the right-hand side of the preceding inequality. Then

$$L(\lambda_i) = -\lambda_i q_i + \log(1 - q_i + q_i e^{\lambda_i}).$$

Prove that $L''(\lambda_i) \leq \frac{1}{4}$. Using a Taylor-MacLaurin expansion, deduce that

$$L(\lambda_i) \leq L(0) + \lambda_i L'(0) + \frac{1}{8}\lambda_i^2 = \frac{1}{8}\lambda^2(b_i - a_i)^2.$$

4) Finally, prove that $\mathbb{P}[S_l \geq (p+t)l] \leq \exp\left(-\lambda l t + \frac{1}{8}\lambda^2 \sum_{i=1}^{l}(b_i - a_i)^2\right)$. Choose the best value for $\lambda > 0$ to deduce the expected inequality.

4.3.4 Generating Function

We call generating function of a random variable X with mean μ the function $M :$ $\mathbb{R} \to \mathbb{R}^+$ defined by $M(t) = \mathbb{E}\left[e^{tX}\right]$. In all that follows, we assume for all considered random variables that $M(t)$ is bounded on an open interval $I \subset \mathbb{R}$, containing 0.

1) Prove that for all x real, $|x|^k \leq \varepsilon^{-k} k! \left(e^{\varepsilon x} + e^{-\varepsilon x}\right)$. Deduce that X has moments $\mathbb{E}\left[X^k\right]$ of any order.

2) Use the asymptotic expansion of the exponential function to prove that

$$M(t) = \sum_{k=0}^{\infty} \frac{\mathbb{E}\left[X^k\right]}{k!} t^k$$

on I and that

$$\mathbb{E}[X] = M'(0), \ \ \mathbb{E}\left[X^k\right] = M^{(k)}(0), \ \ M'(t) = \mathbb{E}\left[X e^{tX}\right], \ \ M''(t) = \mathbb{E}\left[X^2 e^{tX}\right].$$

Hint: To exchange sum and integral by bounded convergence theorem (Lebesgue theorem), use the fact that

$$\sum_{k=1}^{n} \left((t|X|)^k\right)/k! \leq e^{|tX|} \leq e^{tX} + e^{-tX}.$$

3) We set $\Lambda(t) = \log M(t)$. Prove that

$$\Lambda(0) = \log M(0) = 0, \ \ \Lambda'(0) = \frac{M'(0)}{M(0)} = \mu.$$

4) Show that $\Lambda(t)$ is convex on I. More precisely, prove by the Cauchy-Schwartz inequality that

$$\Lambda''(t) = \frac{M(t)M''(t) - M'(t)^2}{M(t)^2} \geq 0.$$

5) Cauchy-Schwartz's inequality $\left(\mathbb{E}[XY]^2 \leq \mathbb{E}\left[X^2\right]\mathbb{E}\left[Y^2\right]\right)$ is an equality if and only if there is a $\lambda \in \mathbb{R}$ such that $X = \lambda Y$. Deduce from this fact that if X has nonzero variance, then $\Lambda(t)$ is strictly convex on I.

6) The Fenchel-Legendre transform of Λ is defined by

$$\Lambda^*(a) = \sup_{t \in \mathbb{R}} \left(at - \Lambda(t)\right), \ \ a \in \mathbb{R}.$$

Prove that Λ^* is convex.

4.3.5 Large Deviations Estimate

This section is devoted to the detailed proof of the following large deviation theorem. The exercise exactly follows the notations and the steps of the proof given in the Grimmett and Stirzaker book [GS01].

Theorem 2 Let $X_1, X_2,...$ be i.i.d. random variables with mean μ and such that $M(t) = \mathbb{E}\left[e^{tX}\right]$ is finite on an open interval containing zero. We set $S_n = X_1 + \cdots + X_n$. Let $a > \mu$ be such that $\mathbb{P}[X > a] > 0$. Then $\Lambda^*(a) > 0$ and

$$-\frac{1}{n}\log\mathbb{P}[S_n > an] \to \Lambda^*(a) \quad \text{when } n \to \infty.$$

Comments: The large number law tells us that $\frac{1}{n}S_n \to \mu$ almost surely. The Central Limit Theorem asserts that typical deviations of $S_n - n\mu$ must be of order $c\sqrt{n}$. Thus, the probabilities $\mathbb{P}[S_n - n\mu] \geq n^\alpha)$ with $\alpha > \frac{1}{2}$ must be small and the preceding theorem yields a precise asymptotic estimate of this smallness when $\alpha = 1$.

As a preliminary question, let us examine the meaning of this theorem when X_i are Bernoulli with parameter $p = \mu$. In that case, $\mathbb{P}[S_n > an] = \mathcal{B}(n, an, p)$ and $M(t) = \mathbb{E}\left[e^{tX}\right] = pe^t + (1-p)$. Thus, $\Lambda(t) = \log(pe^t + (1-p))$. Now,

$$\arg\max_{t>0}(at - \Lambda(t)) = \log\frac{q}{p} + \log\frac{a}{1-a}$$

and, finally,

$$\Lambda^*(a) = a\log\frac{a}{p} + (1-a)\log\frac{1-a}{q}.$$

So the large deviation theorem yields

$$-\frac{1}{n}\log\mathbb{P}(S_n > an) \to a\log\frac{a}{p} + (1-a)\log\frac{1-a}{1-p} \quad \text{when } n \to \infty.$$

Thus, for the binomial law, Hoeffding's inequality and the large deviation behavior yield exactly the same estimate. In this particular framework, Hoeffding's inequality is also known as Bernstein's inequality.

FIRST PART

1) Prove that we can take $\mu = 0$ without loss of generality.
Hint: Assume that the theorem is proved in that particular case and replace X by $X_i - \mu$.

2) By changing X_i into $-X_i$ and by applying the theorem, give the limit of $\frac{1}{n}\log \mathbb{P}[S_n < na]$, for $a < \mu$.

3) Let us prove that $\Lambda^*(a) > 0$. Let $\sigma^2 = \text{var}(X)$. Prove that

$$at - \Lambda(t) = \log\left(\frac{e^{at}}{M(t)}\right) = \log\frac{1 + at + o(t)}{1 + \frac{1}{2}\sigma^2 t^2 + o(t^2)}$$

for t small and deduce that $\Lambda^*(a) > 0$.

4) Use $\Lambda'(0) = \mathbb{E}[X] = 0$ and get

$$\Lambda^*(a) = \sup_{t>0}(at - \Lambda(t)), \quad a > 0.$$

5) In this question and the following ones, one proceeds as for Hoefdding's inequality. Remark that $e^{tS_n} > e^{nat}\mathbb{1}_{S_n>na}$ and derive

$$\mathbb{P}[S_n > na] \le e^{-n(at - \Lambda(t))}.$$

6) Deduce that

$$\frac{1}{n}\log\mathbb{P}[S_n > na] \le -\sup_{t>0}(at - \Lambda(t)) = -\Lambda^*(a).$$

This upper bound brings us halfway to the proof. Now we need a lower bound.

SECOND PART

In this part, we make an assumption on Λ^* which is in practice always satisfied.

Definition 3. *We will say that we stand in the "regular case" at a if the maximum value defining $\Lambda^*(a)$ is attained at a point t in the interior of I, the definition interval for the generating function M. We denote by $\tau = \tau(a)$ this value of t.*

Let us also define $T = \sup\{t, M(t) < \infty.\}$.

1) Check that

$$\Lambda^*(a) = a\tau - \Lambda(\tau), \quad \Lambda'(\tau) = a.$$

2) We set $F(u) = \mathbb{P}[X \le u]$, the distribution function of X. We will associate with this function an auxiliary function defined as follows. Set $d\tilde{F}(u) = \frac{e^{\tau u}}{M(\tau)}dF(u)$, or, in other terms,

$$\tilde{F}(y) = \frac{1}{M(\tau)}\int_{-\infty}^{y} e^{\tau u}\,dF(u).$$

Consider $\tilde{X}_1, \tilde{X}_2, \ldots$ independent random variables all having $\tilde{F}$ as distribution function and set $\tilde{S}_n = \tilde{X}_1 + \cdots + \tilde{X}_n$.

3) By using the general formula $M(t) = \mathbb{E}[e^{tX}] = \int_{\mathbb{R}} e^{tu}\,dF(u)$, prove that

$$\tilde{M}(t) = \frac{M(t+\tau)}{M(\tau)}.$$

4) Deduce that

$$\mathbb{E}[\tilde{X}_i] = \tilde{M}'(0) = a,$$

$$\mathrm{var}(\tilde{X}_i) = \mathbb{E}[\tilde{X}_i^2] - \mathbb{E}[\tilde{X}_i]^2 = \tilde{M}''(0) - \tilde{M}'(0)^2 = \Lambda''(\tau).$$

5) Prove that the generating function of $\tilde{S}_n$ is

$$\left(\frac{M(t+\tau)}{M(\tau)}\right)^n = \frac{1}{M(\tau)^n}\int_{\mathbb{R}} e^{(t+\tau)u}dF_n(u),$$

where F_n is the distribution function of S_n. Deduce that the distribution function $\tilde{F}_n$ of $\tilde{S}_n$ satisfies

$$d\tilde{F}_n(u) = \frac{e^{\tau u}}{M(\tau)^n}dF_n(u).$$

6) Let $b > a$. Deduce from the preceding question that

$$\mathbb{P}[S_n > na] \geq e^{-n(\tau b - \Lambda(\tau))}\mathbb{P}[na < \tilde{S}_n < nb].$$

7) By using the fact that the mean of $\tilde{X}_i$ is a, the law of large numbers, and the Central Limit Theorem, prove that $\mathbb{P}[na < \tilde{S}_n < nb] \to \frac{1}{2}$. Deduce from this fact and from the preceding question that

$$\liminf_{n\to\infty} \frac{1}{n}\log\mathbb{P}[S_n > na] \geq -(\tau b - \Lambda(\tau)) \to -\Lambda^*(a) \quad \text{when } b \to a.$$

THIRD PART

1) We make the same assumptions as in the preceding part, but we are now in the nonregular case. In order to get back to the regular case, we fix $c > a$ and we set $X^c = \inf(X, c)$. Denote by $M^c(t)$ the associated generating function and set $\Lambda^c(t) = \log M^c(t)$.

2) Prove that $M^c(t) < \infty$ for every $t > 0$ and that $\mathbb{E}[X^c] \to 0$ when $c \to \infty$ and $\mathbb{E}[X^c] \leq 0$.

3) Show that there exists $b \in (a, c)$ such that $\mathbb{P}[X > b] > 0$ and deduce that

$$at - \Lambda^c(t) \leq at - \log(e^{tb}\mathbb{P}[X > b]) \to -\infty \quad \text{when } t \to \infty.$$

4) Conclude that the sequence X_i^c stands in the "regular case" –in other terms that the supremum of $(at - \Lambda^c(t))$ for $t > 0$ is attained at a value $\tau^c \in (0, \infty)$– and, finally, that

$$\frac{1}{n}\log\mathbb{P}\left[\sum_{i=1}^n X_i^c > na\right] \to -(\Lambda^c)^*(a) \quad \text{when } t \to \infty.$$

(We set $\Lambda^{c*} = \sup_{t>0}(at - \Lambda^c(t)) = a\tau^c - \Lambda^c(\tau^c)$.)

5) Prove that $\Lambda^{c*}(a) \downarrow \Lambda^{\infty*}$ when $c \to +\infty$. and $0 \leq \Lambda^{\infty*} < \infty$.

6) Prove that

$$\frac{1}{n}\log\mathbb{P}[S_n > na] \geq \frac{1}{n}\log\mathbb{P}\left[\sum_{i=1}^n X_i^c > na\right].$$

7) Explain why it suffices to prove that $\Lambda^{\infty*} \leq \Lambda^*(a)$ in order to conclude the proof of the theorem in the irregular case.

8) To show this last relation, prove that the set $I_c = \{t \geq 0, \; at - \Lambda^c(t) \geq \Lambda^{\infty*}\}$ is not empty. Prove that I_c is a compact interval. Prove that the intervals I_c decrease (by inclusion) as c grows and deduce that there is some $\zeta \in \cap_{c>a} I_c$. Finally, show that $\Lambda^c(\zeta) \to \Lambda(\zeta)$ when $c \to \infty$ and that $a\zeta - \Lambda(\zeta) = \lim_{c\to\infty}(a\zeta - \Lambda^c(\zeta)) \geq \Lambda^{\infty*}$. Conclude the proof.

4.3.6 The Central Limit Theorem

4.3.6.1 Some Basics on the Characteristic Function of a Random Variable and on the Gauss Function

Let X be a random variable with values in $\mathbb{R}^d$. For $t \in \mathbb{R}^d$, let

$$\varphi_X(t) = \mathbb{E}\left[e^{it.X}\right] = \int_{\mathbb{R}^d} e^{it.x} d\mu_X(dx) = \int_{\mathbb{R}^d} e^{it.x} f_X(x)\, dx$$

be its characteristic function where it is assumed that the random variable X has a density $f_X(t)$ with respect to the Lebesgue measure and $t \cdot x = <t,x> = \sum_{i=1}^d t_i x_i$ denotes the scalar product of t and x.

Briefly prove the following properties (φ is written for φ_X):

1. $\varphi(0) = 1$.
2. $\forall t, |\varphi(t)| \leq 1$.
3. $t \to \varphi(t)$ is continuous (use Lebesgue dominated convergence theorem).
4. $\varphi_{(-X)}(t) = \overline{\varphi_X(t)}$.
5. $\varphi_{aX+b}(t) = e^{ib.t}\varphi_X(at), \quad (a \in \mathbb{R}, \; b \in \mathbb{R}^d)$.
6. If X and Y are independent, $\varphi_{X+Y} = \varphi_X \varphi_Y$.

Prove the following properties of Gaussian distributions:

1. If X is Gaussian and has density $\frac{1}{\sigma\sqrt{2\pi}} e^{-\frac{(x-\mu)^2}{2\sigma^2}}$, then $\varphi_X(t) = e^{i\mu t - \frac{\sigma^2 t^2}{2}}$.

   Hint: Prove this result first when $\mu = 0$ and $\sigma = 1$. Set $\varphi(t) = \frac{1}{\sqrt{2\pi}} \int e^{-itx} e^{-\frac{x^2}{2}} dx$ and prove that $\varphi'(t) = t\varphi(t)$, $\varphi(0) = 1$. It will follow that $\varphi(t) = e^{-\frac{t^2}{2}}$.
2. The variance of X is σ^2 and its mean is μ.
3. If X_1 and X_2 are Gaussian independent, then $X_1 + X_2$ also is Gaussian: Compute the mean and variance of $X_1 + X_2$ as a function of those (μ_i, σ_i^2), $i = 1$, 2, of X_1 and X_2.

4.3.6.2 Asymptotic Expansion of the Characteristic Function

Let X be a random variable with values in $\mathbb{R}^d$ and admitting an order 2 moment (i.e., $\mathbb{E}\left[X^2\right] = \int_{\mathbb{R}^d} x^2 d\mu_X(x) < \infty$).

1)Prove that for every real number x,

$$\exp(ix) = 1 + ix - x^2 \int_0^1 (1-u)\exp(iux)\,du.$$

2) Deduce that

$$\left|\exp(ix) - \left(1 + ix - x^2/2\right)\right| \le x^2.$$

3) By applying the method of the two preceding questions, prove that

$$\left|\exp(ix) - \left(1 + ix - x^2/2\right)\right| \le |x|^3/6.$$

4) Deduce that

$$\left|\exp(ix) - \left(1 + ix - x^2/2\right)\right| \le \inf(x^2, |x|^3/6)$$

and

$$\left|\varphi_X(t) - \left(1 + it\mathbb{E}[X] - \frac{t^2}{2}\mathbb{E}\left[X^2\right]\right)\right| \le t^2\mathbb{E}\left[\inf(X^2, |t|\|X\|^3/6)\right].$$

By applying the Lebesgue theorem to the right-hand term, deduce an asymptotic expansion of order 2 of $\varphi_X(t)$ at $t = 0$.

4.3.6.3 Central Limit Theorem

We shall prove the following theorem (Central limit theorem in $\mathbb{R}^d$):

Theorem 3 Let $(X_n)_{n\ge 1}$ be a sequence of random variables defined on the same probability space $(\Omega, \mathcal{F}, \mathbb{P})$, with values in $\mathbb{R}^d$. We assume that they are independent and identically distributed and that $\mathbb{E}\left[\|X_i\|^2\right] < +\infty$. Then the sequence Y_n defined by

$$Y_n = \frac{1}{\sqrt{n}}\sum_{j=1}^n (X_j - \mathbb{E}[X_j])$$

converges in law toward a d-dimensional Gaussian distribution with zero mean and whose covariance matrix is the same as the one of the X_i's.

We recall that a sequence of random variables is said to converge in law toward a random variable X if for every continuous and bounded function g on $\mathbb{R}^d$, one has

$$\int_{\mathbb{R}^d} g(x)d\,\mu_{X_n}(x) \longrightarrow \int_{\mathbb{R}^d} g(x)d\,\mu_X(x).$$

We will, however, use a convergence in law criterion, Lévy's theorem, whose statement is given in question 5.

1) Prove that the characteristic function of Y_n is given by:

$$\varphi_{Y_n}(t) = \left[\varphi_{<X_1 - \mathbb{E}[X_1],t>}\left(\frac{1}{\sqrt{n}}\right)\right]^n.$$

2) By using the results we just proved on the asymptotic expansion of the characteristic function, deduce that

$$\varphi_{Y_n}(t) = \left[1 - \frac{1}{2n}\mathbb{E}\left[< X_1 - \mathbb{E}[X_1], t >^2\right] + o\left(\frac{1}{n}\right)\right]^n.$$

3) Prove that for every $z \in \mathbb{C}$,

$$\lim_{n \to +\infty}\left(1 + \frac{z}{n}\right)^n = \exp(z),$$

the convergence being uniform on every bounded set. (Use the complex logarithm and an asymptotic expansion of order 1.)

Detailed hint for question 3: The principal determination of the logarithm, denoted by $\log(z)$, can be used. It satisfies $e^{\log(z)} = z$ if $|z - 1| < 1$ and its asymptotic expansion yields $\log(1 + z) = z - \frac{1}{2}z^2 + \frac{1}{3}z^3 + \cdots$. Thus, for $|z| \leq A$ and for n large enough $(n > A)$,

$$\left|\log\left(1 + \frac{z}{n}\right)^n - z\right| = \left|-\frac{z^2}{2n} + \frac{z^3}{3n^2} - \frac{z^4}{4n^3} + \cdots\right| \leq \frac{A^2}{n}\left(\frac{1}{2} + \frac{A}{3n} + \frac{A^2}{4n^2} + \cdots\right) \to 0$$

when $n \to \infty$. Thus, $\log(1 + \frac{z}{n})^n \to z$ uniformly for $|z| < A$ when $n \to \infty$. From this inequality also follows that $|\log(1 + \frac{z}{n})^n| \leq A + 1$ for n large enough. The exponential function being uniformly continuous on the set $\{|z| \leq A + 1\}$, the result follows.

4) Deduce that

$$\lim_{n \to +\infty}\varphi_{Y_n}(t) = \exp\left[-\frac{1}{2} < Ct, t >\right],$$

where C is the covariance matrix of X_1.

5) Conclude by applying Lévy's theorem: If X_n are real random variables such that $\varphi_{X_n} \to \varphi$ and if φ is a continuous function at 0, then there exists a random variable X such that $\varphi_X = \varphi$ and the sequence X_n converges in law to X.

4.3.7 The Tail of the Gaussian Law

In this exercise, we set $\phi(y) = (2\pi)^{-1/2}\exp(-y^2/2)$ and $\Phi(x) = \int_{-\infty}^{x}\phi(y)\,dy$. We consider a random variable X with density law ϕ.

1) Remark that $\phi'(y) = -y\phi(y)$ and deduce that for every $x > 0$,

$$x^{-1}\phi(x) \geq (1 - \Phi(x)).$$

2) Remark that $(y^{-1}\phi(y))' = -(1 + y^{-2})\phi(y)$ and show that for every $x > 0$,

$$\phi(x) \leq (x + x^{-1})(1 - \Phi(x)).$$

3) Deduce upper and lower bounds for $\mathbb{P}(X > x)$ for $x > 0$.

Chapter 5
Alignments in Digital Images

Digital images usually have many alignments due to perspective, human made objects, and so forth. Can they and only they be detected? There should be *detection thresholds* telling us whether a particular configuration of points is aligned enough to pop out as an alignment. Alignments are not trivial events. They depend *a priori* on four different parameters, namely the total length l of the alignment, the number k of observed aligned points in it, the precision p of the alignment, and the size of the image N. So a decision threshold function $k_{min}(l,p,N)$ is needed and will be established by the Helmholtz principle. In its weak formulation, this principle commonsensically formulates that k_{min} should be fixed in such a way as to seldom detect any alignment in a white noise image. In the stronger formulation, the Helmholtz principle tells us that whenever a configuration occurs, which could not arise by chance in white noise, this configuration is perceived and must be detected by a Computer Vision algorithm. In Section 5.1, we define and analyze the white noise image *a contrario* and show how to compute detection thresholds $k_{min}(l,p,N)$ discarding alignments in white noise. Section 5.2 is devoted to the analysis of the *Number of False Alarms (NFA)* of an alignment and the rest of the chapter considers several estimates of the detection threshold k_{min}. Finally, several consistency problems associated with the definition of meaningfulness are considered. In Section 5.5, the important problem of choosing the precision p is finally addressed.

5.1 Definition of Meaningful Segments

Alignments in digital images are usually not dot alignments but correspond to segments where the gradient is observed to be roughly orthogonal to the segment's direction. The first question is how to compute the gradient and how to sample the segments (how many different points can be counted on a digital segment?).

5.1.1 The Discrete Nature of Applied Geometry

Perceptual and digital images are the result of a convolution followed by a spatial sampling, as described in the Shannon-Whittaker theory. From the samples, a continuous image may be recovered by Shannon interpolation, but the samples by themselves contain all of the image information. From this point of view, one could claim that no absolute geometric structure is present in an image, (e.g. no straight line, no circle, no convex set, etc). We claim in fact the opposite and the following definition will explain how we can be sure that a line is present in a digital image.

Consider a gray-level image of size N (i.e. a regular grid of N^2 pixels). At each point x, or pixel, of the discrete grid, there is a gray level $u(x)$ that is quantized and therefore inaccurate. We may compute at each point the direction of the gradient, which is the simplest local contrast invariant information (local contrast invariance is a necessary requirement in image analysis and perception theory [Met75], [Wer23]). The direction of the level line passing by the point can be calculated on a $q \times q$ pixels neighborhood (generally, $q = 2$).

The computation of the gradient is based on an interpolation (we have $q = 2$). We define the *direction* at pixel (n,m) by rotating by $\pi/2$ the direction of the gradient of the order 2 interpolation at the center of the 2×2 window made of pixels (n,m), $(n+1,m)$, $(n,m+1)$, and $(n+1,m+1)$. We get

$$\mathrm{dir}(n,m) = \frac{1}{\| Du(n,m) \|} Du(n,m)^{\perp}, \tag{5.1}$$

where

$$Du(n,m) = \frac{1}{2} \left(\begin{array}{c} [u(n+1,m)+u(n+1,m+1)] + [u(n,m)+u(n,m+1)] \\ [u(n,m+1)+u(n+1,m+1)] - [u(n,m)+u(n+1,m)] \end{array} \right). \tag{5.2}$$

Formula (5.2) can be interpreted as the exact gradient $D\tilde{u}(n+1/2,m+1/2)$ when $\tilde{u}$ is the bilinear interpolate of u. This interpolate is defined in $[n,n+1] \times [m,m+1]$ by

$$\tilde{u}(x,y) = (y-m) \Big((x-n)X_4 + (1-x+n)X_3 \Big)$$

$$+(1-y+m) \Big((x-n)X_2 + (1-x+n)X_1 \Big),$$

where $X_1 = u(n,m)$, $X_2 = u(n+1,m)$, $X_3 = u(n,m+1)$, and $X_4 = u(n+1,m+1)$.

We say that two points X and Y have the same direction with precision $1/n$ if

$$|\mathrm{Angle}(\mathrm{dir}(X),\mathrm{dir}(Y))| \leq \frac{\pi}{n}. \tag{5.3}$$

In agreement with psychophysics and numerical experimentation, realistic values of n range from 32 to 4 and it is, in general, useless to consider larger values of n. Nothing hinders, however, the consideration of larger n's for very well-sampled images for special scientific or technical applications.

5.1.2 The A Contrario Noise Image

In line with the Helmholtz principle, an alignment in a digital image is meaningful if it could not happen in a same size white noise image. Following Attneave [Att54] and Shannon [Sha48], an image is a "white noise" if the values at each pixel are independent and identically distributed random variables. The joint distribution of the gray-level at each pixel is ideally Gaussian or uniform. Actually, the following proposition proves that the direction at each point in the noise image is a uniformly distributed random variable on $[0, 2\pi]$ when the gray level distribution is Gaussian. It is also nearly uniform when the gray-level distribution is uniform.

Proposition 8 *Let u be a gray-level image. Let Du denote the gradient of u computed according to Equation (5.2). We use complex numbers notations and write $Du = R\exp(i\theta)$, where $R = |Du|$ is the modulus of Du and θ its argument. Then the following hold:*

1. *If u is a Gaussian white noise, which means that the gray levels at all of the pixels are independent and identically Gaussian $\mathcal{N}(\mu, \sigma^2)$ distributed, then θ is uniformly distributed on $[0, 2\pi]$.*
2. *If u is a uniform white noise, which means that the gray levels at all of the pixels are independent and identically uniformly distributed on $[-\frac{1}{2}, \frac{1}{2}]$, then the law of θ is given by the density function g, $\pi/2$-periodic and whose restriction to $[-\pi/4, \pi/4]$ is*

$$g(\theta) = \frac{1}{12}\left(1 + \tan^2\left(\frac{\pi}{4} - |\theta|\right)\right)\left(2 - \tan\left(\frac{\pi}{4} - |\theta|\right)\right)$$

(see Figure 5.1).

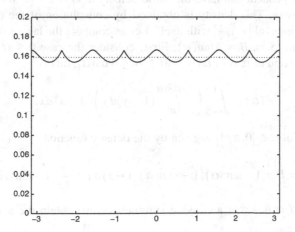

Fig. 5.1 Law of θ on $[-\pi, \pi]$ when the image is a uniform noise, and comparison with the uniform distribution on $[-\pi, \pi]$ (dotted line).

Proof — According to Equation (5.2), the gradient Du at a pixel (n, m) is defined by

$$Du(n,m) = \begin{pmatrix} u_x \\ u_y \end{pmatrix} := \frac{1}{2} \begin{pmatrix} X_2 + X_4 - X_1 - X_3 \\ X_1 + X_2 - X_3 - X_4 \end{pmatrix}, \tag{5.4}$$

where $X_1 = u(n,m)$, $X_2 = u(n+1,m)$, $X_3 = u(n,m+1)$, and $X_4 = u(n+1,m+1)$.

Set $A = X_2 - X_3$ and $B = X_1 - X_4$. We have $u_x = \frac{A-B}{2}$ and $u_y = \frac{A+B}{2}$. Since $Du = u_x + iu_y = R\exp(i\theta)$, we also have

$$A + iB = R\sqrt{2} \cdot \exp\left[i(\theta - \pi/4)\right]. \tag{5.5}$$

Let us assume here that X_1, X_2, X_3, and X_4 are independent random variables with the same Normal distribution $\mathcal{N}(\mu, \sigma^2)$. Then A and B are independent and have both the same Gaussian distribution of mean 0 and variance $2\sigma^2$. The law of the couple (A, B) is given by the density function

$$f(a, b) = \frac{1}{4\pi\sigma^2} \exp\left(-\frac{a^2 + b^2}{4\sigma^2}\right).$$

Thus, using the polar coordinates, the density of the couple (R, θ) is

$$h(r, \theta) = \frac{r}{4\pi\sigma^2} \exp\left(-\frac{r^2}{4\sigma^2}\right) \mathbb{1}_{[0, 2\pi]}(\theta),$$

which shows in particular that θ is almost surely defined. It also shows that R and θ are independent and the law of θ is uniform for every given $R = r$. This proves the first part of the proposition.

Let us now assume that X_1, X_2, X_3, and X_4 are independent random variables uniformly distributed on $[-\frac{1}{2}, \frac{1}{2}]$. Then the random variables $A = X_2 - X_3$ and $B = X_1 - X_4$ are independent and have the same density $h(x) = 1 - |x|$ for $|x| \leq 1$, and $h(x) = 0$ otherwise. This density is obtained by convolution of the characteristic function of the interval $[-\frac{1}{2}, \frac{1}{2}]$ with itself. Let us compute the law of $\alpha = \theta - \pi/4$. Due to Equation (5.5), $B = A\tan(\alpha)$. First, consider the case $0 \leq \alpha \leq \pi/4$. The distribution function of α is $F(\alpha) = \mathbb{P}[0 \leq B \leq A\tan\alpha]$; that is,

$$F(\alpha) = \int_{x=0}^{1} \left(\int_{y=0}^{x\tan\alpha} (1-y)dy\right)(1-x)\,dx.$$

Hence, the law of $\alpha \in [0, \pi/4]$ is given by the density function

$$f(\alpha) = F'(\alpha) = \int_0^1 x(1+\tan^2\alpha)(1 - x\tan\alpha)(1-x)\,dx = \frac{1}{12}(1+\tan^2\alpha)(2 - \tan\alpha).$$

Finally, since $\alpha = \theta - \pi/4$ and using symmetries, we obtain the announced law for θ. $\square$

The histogram of the gradient orientation is very sensitive to quantization of the gray levels of the image. Think of the extreme case of a binary image u, where the

(a) White noise: no alignment detected ($\varepsilon = 1$)   (b) Convolution of image (a) with a Gaussian kernel with standard deviation of 4 pixels: no alignment detected ($\varepsilon = 1$)

Fig. 5.2 The weak Helmholtz principle: "no detection in noise". According to Attneave [Att54] and Shannon [Sha48], an image is a white noise if the values at each pixel are independent and identically distributed random variables. No geometric structure arises in such images at first sight – in particular, no alignment. This demonstrates experimentally the validity of the Helmholtz principle. The left-hand image is a white noise. The right-hand image is a white noise image blurred by the convolution with a Gaussian of 4 pixels size standard deviation. The alignment detector defined in the present chapter did not detect any 1-meaningful alignment in these images. This algorithm is actually designed to detect at most one alignment on average in white noise.

gradient orientation θ computed by formulas (5.1)–(5.3) takes only eight values, all multiples of $\pi/4$. Digital images usually contain many flat regions and, in these regions, gray levels take a few quantized values. Consequently, the gradient orientation is very quantized and will thus be responsible for false alignment detection. In order to avoid this problem, a dequantization has to be performed. One possible dequantization method (a Shannon $(\frac{1}{2}, \frac{1}{2})$-translation of the image) has been proposed and studied in [DLMM02].

In such images as the ones presented in Figure 5.2, no geometric structure can be seen and, in particular, no alignment. This illustrates the validity of the Helmholtz principle in its weak form. Applying the stronger form, we will detect an alignment in a digital image *if and only if it could hardly occur in a white noise image of the same size*. Of course, if we do many experiments, we will end up with a noise image showing some geometric structure and the same occurs if we take a very large noise image. In the former informal definition, it is therefore important to mention that the a-contrario noise image has the *same size* as the original image.

In the following, we assume that the accuracy parameter n is larger than 2 and we set $p = \frac{1}{n} < \frac{1}{2}$; p is the accuracy of the direction. We can interpret p as the probability that two independent points have the same direction with the given accuracy p. In a white noise image, two pixels at a distance larger than 2 have independent directions when this direction is computed by formulas (5.1)–(5.3) and, according to Proposition 8, this direction is uniformly distributed on $[0, 2\pi]$. Let A be a segment in the a-contrario noise image made of l independent pixels. This means that the distance between two consecutive points of A is 2 and so the length of A is $2l$. We are interested in the number of points of A having their direction aligned with the direction of A for the precision p. Such points of A will simply be called *aligned points of A*.

Our aim is to compute the minimal number $k(l)$ of aligned points that must be observed on a length l segment to make the alignment meaningful.

5.1.3 Meaningful Segments

Let A be a straight segment with length l and $x_1, x_2, \ldots, x_l$ be the l (independent) points of A. Let X_i be the random variable whose value is 1 when the direction at pixel x_i is aligned with the direction of A for the precision p, and 0 otherwise. We then have the following Bernoulli distribution for X_i:

$$\mathbb{P}[X_i = 1] = p \quad \text{and} \quad \mathbb{P}[X_i = 0] = 1 - p.$$

The random variable representing the number of x_i having the right direction is

$$S_l = X_1 + X_2 + \cdots + X_l.$$

Because of the independence of the X_i's, the law of S_l is given by the binomial distribution

$$\mathbb{P}[S_l = k] = \binom{l}{k} p^k (1 - p)^{l-k}.$$

Given a length l segment we want to know whether it is ε-meaningful among all of the segments of the image (not just among the segments having the same length l). Let $m(l)$ be the number of oriented segments of length l in the $N \times N$ noise image. Define the total number of oriented segments in a $N \times N$ image as the number of couples (x, y) of points in the image (an oriented segment is given by its starting point and its ending point) and thus we have

$$\sum_{l=1}^{l_{max}} m(l) = N^2(N^2 - 1) \simeq N^4.$$

For the sake of simplicity, the value N^4 is kept as an order of magnitude for the number of segments. All segments in the $N \times N$ image are numbered from $i = 1$ to $i = N^4$.

Definition 4 (Detection thresholds). *We call* detection thresholds *a family of positive values* $w(l, \varepsilon, N)$, $1 \leq l \leq l_{max}$, *such that*

$$\sum_{l=1}^{l_{max}} w(l, \varepsilon, N) m(l) \leq \varepsilon.$$

Definition 5 (General definition of an ε-meaningful segment). *A length l segment is ε-meaningful in a N × N image if it contains at least k(l) points having their direction aligned with the one of the segment, where k(l) is given by*

$$k(l) = \min\left\{k \in \mathbb{N}, \ \mathbb{P}[S_l \geq k] \leq w(l, \varepsilon, N)\right\}.$$

Let us develop and explain this definition. For $1 \leq i \leq N^4$, let e_i be the event "the i-th segment is ε-meaningful" and χ_{e_i} denote the characteristic function of the event e_i. We have

$$\mathbb{P}[\chi_{e_i} = 1] = \mathbb{P}\left[S_{l_i} \geq k(l_i)\right],$$

where l_i is the length of the i-th segment. Notice that if l_i is small, we may have $\mathbb{P}\left[S_{l_i} \geq k(l_i)\right] = 0$. Let R be the random variable representing the exact number of e_i occurring simultaneously in a trial. Since $R = \chi_{e_1} + \chi_{e_2} + \cdots + \chi_{e_{N^4}}$, the expectation of R is

$$\mathbb{E}[R] = \mathbb{E}[\chi_{e_1}] + \mathbb{E}[\chi_{e_2}] + \cdots + \mathbb{E}\left[\chi_{e_{N^4}}\right] = \sum_{l=0}^{l_{max}} m(l)\mathbb{P}[S_l \geq k(l)].$$

We compute here the expectation of R but not its law because it depends a lot on the relations of dependence between the e_i. The main point is that segments may intersect and overlap, so that the e_i events are not independent and may even be strongly dependent.

By definition we have

$$\mathbb{P}[S_l \geq k(l)] \leq w(l, \varepsilon, N),$$

so that

$$\mathbb{E}[R] \leq \sum_{l=1}^{l_{max}} w(l, \varepsilon, N)m(l) \leq \varepsilon. \tag{5.6}$$

This means that the expectation of the number of ε-meaningful segments in an image is less than ε. The question now is how to set the detection thresholds. The number of discrete segments with length l in a digital $N \times N$ image has N^3 order of magnitude. There are indeed approximately N^2 possible discrete straight lines in a $N \times N$ image, and on each discrete line there are about N choices for the starting point of the segment. So $m(l) \simeq N^3$. If we were interested in segments with length l only, we should take $w(l, \varepsilon, N) = \frac{\varepsilon}{N^3}$. There is no reason to be more interested in large segments than in short ones. Thus, a uniform weighting over lengths is sound. Our final definition of an ε-meaningful segment will therefore admit this uniformity principle:

$$\forall l \geq 1, \quad w(l, \varepsilon, N) = \frac{\varepsilon}{N^4}.$$

This leads to the final simpler definition of ε-meaningful segments.

Definition 6 (ε-meaningful segment: final definition). *A length l segment is ε-meaningful in a N × N image if it contains at least k(l) points having their direction aligned with the one of the segment, where k(l) is given by*

$$k(l) = \min\left\{k \in \mathbb{N}, \ \mathbb{P}[S_l \geq k] \leq \frac{\varepsilon}{N^4}\right\}. \tag{5.7}$$

(a) The original image.

(b) Maximal ε-meaningful segments, for $\varepsilon = 1/10$.

(c) Maximal ε-meaningful segments, for $\varepsilon = 1/100$.

(d) Maximal ε-meaningful segments, for $\varepsilon = 1/1000$.

Fig. 5.3 Airport image (size 510×341). This digital image has a noticeable aliasing distortion creating horizontal and vertical dashes along the edges. We display in image (b) all maximal meaningful segments for $\varepsilon = 1/10$ (as we already mentioned, maximal meaningful segments will be defined in Chapter 6; they are in some sense the "best representatives" of the set of all meaningful segments). Images (c) and (d) show the result of the same experiment for $\varepsilon = 1/100$ and $\varepsilon = 1/1000$. The very similar results illustrate the $\log \varepsilon$ dependence of detection: We get almost the same detections when changing ε by one or two orders of magnitude. This stability is all the more true because alignments in a digital image use to be much longer than needed to be detected.

In the following, we write $\mathcal{B}(l,k)$ for $\mathcal{B}(l,k,p) = \mathbb{P}[S_l \geq k]$; that is, we omit mentioning p in the arguments of $\mathcal{B}$ when p is fixed.

First examples of meaningful segments in an image are displayed and commented in Figures 5.3 and 5.4. These figures display first all meaningful segments and then a more accurate selection, the maximal meaningful segments. This selection process will be described in Chapter 6.

5.1.4 Detectability Weights and Underlying Principles

Before proceeding with the properties of meaningful segments, we should further discuss the choice of "detection thresholds". Consider a fixed length l. It is apparent by an obvious translation and rotation invariance principle that all segments with

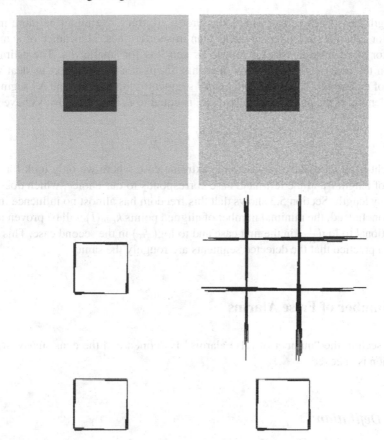

Fig. 5.4 Meaningful segments on a simple example. On the first row, we show on the left the original image of a uniform square on a uniform background (its size is 256×256). On the right, we have a noisy version of this image (degraded by an additive Gaussian noise). The second row contains all the meaningful segments of each image. One can clearly see that, since noise creates pixels with horizontal and vertical orientations in the background and since the boundaries of the square are very meaningful segments, they can be extended in the background while being still meaningful. The third row presents the maximal meaningful segments of each image: They are defined as the segments which have the lowest number of false alarms among the set of segments which are contained in or contain them. These maximal meaningful segments are in some sense the "best representatives" of the set of all meaningful segments. Their definition and their properties will be detailed in Chapter 6. In the experiments we will generally only display maximal meaningful segments since with all meaningful segments, the image is almost entirely covered by segments!

the same orientation should have the same chances to be detected. This is why the above definition of detection thresholds made them depend only on the length l, the image size N, and the allowed expectation of "false alarms", ε. In the final Definition 6 of ε-meaningful segments, the requirement that $w(l, N, \varepsilon)$ is independent of the length l has been added. This looks pretty arbitrary. We might be more interested, for example, in longer segments than in shorter ones. The choice of w might allow for more false alarms for longer l's and less for smaller l's. The estimates given in the next section will show that this discussion is futile. Let us deal with orders of magnitude. There are roughly N^3 segments with length l and N^4 segments of any length. If the global false alarm rate is equal to ε, one must always have

$$\frac{\varepsilon}{N^4} \leq w(l,N,\varepsilon) \leq \frac{\varepsilon}{N^3}.$$

The right-hand case corresponds to the extreme case where we only look for segments of length l and the left-hand case corresponds to our choice, which does not favor any length. Section 5.3 shows that this freedom has almost no influence in the detection. Indeed, the minimal number of aligned points $k_{min}(l)$ will be proven to be proportional to $\log(\frac{\varepsilon}{N^4})$ in the first case and to $\log(\frac{\varepsilon}{N^3})$ in the second case. This will mean in practice that the detected segments are roughly the same.

5.2 Number of False Alarms

In this section the "number of false alarms" is defined and the consistency of this definition is checked.

5.2.1 Definition

Definition 7 (Number of false alarms). *Let A be a segment of length l_0 with k_0 points having their direction aligned with the direction of A. We define the number of false alarms of A as*

$$\text{NFA}(l_0,k_0) = N^4 \cdot \mathbb{P}\left[S_{l_0} \geq k_0\right] = N^4 \cdot \sum_{k=k_0}^{l_0} \binom{l_0}{k} p^k (1-p)^{l_0-k}.$$

Proposition 9 *Let $A = (l_0,k_0)$ be a segment. Then the segment A is ε-meaningful if and only if $\text{NFA}(A) \leq \varepsilon$. In other words, $\text{NFA}(A)$ is the smallest value of ε such that A is ε-meaningful.*

Proof — By definition of $\text{NFA}(A)$, we have $N^4 \cdot \mathbb{P}\left[S_{l_0} \geq k_0\right] = \text{NFA}(l_0,k_0)$ and therefore

$$\mathbb{P}\left[S_{l_0} \geq k_0\right] = \frac{\text{NFA}(l_0,k_0)}{N^4}.$$

By the definition (5.7) of $k(l)$ and since $\mathbb{P}[S_l \geq k]$ is decreasing with respect to k, $k_0 = k(l_0)$. Thus, A is $\text{NFA}(l_0,k_0)$-meaningful and is no more ε-meaningful if $\varepsilon < \text{NFA}(l_0,k_0)$. $\square$

In all of the following, a function called NFA will always satisfy a property like Proposition 9; that is, any event A such that $NFA(A) \leq \varepsilon$ will be ε-meaningful. This property can even be taken as a definition of a NFA, as done recently in [GM06]. This point is discussed in Section 5.6.

5.2.2 Properties of the Number of False Alarms

Proposition 10 *The number of false alarms* $\mathrm{NFA}(l_0, k_0)$ *has the following properties:*

1. $\mathrm{NFA}(l_0, 0) = N^4$, *which proves that the event for a segment to have more than zero aligned points is never meaningful!*
2. $\mathrm{NFA}(l_0, l_0) = N^4 \cdot p^{l_0}$, *which shows that a segment where all points have the "good" direction is ε-meaningful if its length is larger than* $(-4\log N + \log \varepsilon)/ \log p$.
3. $\mathrm{NFA}(l_0, k_0 + 1) < \mathrm{NFA}(l_0, k_0)$. *In other terms, if two segments have the same length l_0, the more meaningful is the one that has the more aligned points.*
4. $\mathrm{NFA}(l_0, k_0) < \mathrm{NFA}(l_0 + 1, k_0)$. *This property can be illustrated by the following figure of a segment (where a $\bullet$ represents a misaligned point and a $\rightarrow$ represents an aligned point):*

$$\rightarrow \rightarrow \bullet \rightarrow \rightarrow \bullet\bullet \rightarrow \rightarrow \rightarrow \rightarrow \rightarrow \bullet$$

If we remove the last point on the right that is misaligned, the new segment is less probable and therefore more meaningful than the considered one.

5. $\mathrm{NFA}(l_0 + 1, k_0 + 1) < \mathrm{NFA}(l_0, k_0)$. *Again, we can illustrate this property:*

$$\rightarrow \rightarrow \bullet \rightarrow \rightarrow \bullet\bullet \rightarrow \rightarrow \rightarrow \rightarrow \rightarrow \rightarrow$$

If we remove the last aligned point (on the right), the new segment is more probable and therefore less meaningful than the considered one.

Proof — This proposition is an easy consequence of the definition and properties of the binomial distribution (see [Fel68]). A detailed proof is given in Exercise 4.3.1 at the end of the chapter. □

Now, consider a length l segment (made of l independent pixels). The expectation of the number of points of the segment having the same direction as the segment is simply the expectation of the random variable S_l

$$\mathbb{E}[S_l] = \sum_{i=1}^{l} \mathbb{E}[X_i] = \sum_{i=1}^{l} \mathbb{P}[X_i = 1] = p \cdot l.$$

The ε-meaningful segments, whose NFA is less than ε, have a probability less than ε/N^4. Since they represent alignments (deviations from randomness), they should contain more aligned points than the expected number computed above. This remark is the main point of the following proposition.

Proposition 11 *Assume that $p \leq \frac{1}{2}$. Let A be a segment of length $l_0 \geq 1$ containing at least k_0 points having the same direction as the one of A. If* $\mathrm{NFA}(l_0, k_0) \leq p \cdot N^4$ *(which is the case when A is 1-meaningful because N is very large and thus, $pN^4 > 1$), then*

$$k_0 \geq pl_0 + (1-p).$$

This is a "sanity check" for the model. This proposition will be proven by Lemma 2.

5.3 Orders of Magnitudes and Asymptotic Estimates

In this section precise asymptotic and nonasymptotic estimates of the thresholds $k(l)$ will be given. They roughly say that

$$k(l) \simeq pl + \sqrt{C \cdot l \cdot \log \frac{N^4}{\varepsilon}},$$

where $2p(1-p) \leq C \leq \frac{1}{2}$. Some of these results are illustrated in Figure 5.5. These estimates are not necessary for the algorithm (because $\mathcal{B}(l,k)$ is easy to compute),

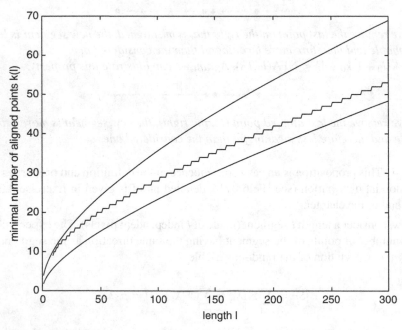

Fig. 5.5 Estimates for the threshold of meaningfulness $k(l)$: The middle (stepcase) curve represents the exact value of the minimal number of aligned points $k(l)$ to be observed on a 1-meaningful segment of length l in an image of size 512, for a direction precision of $1/16$. The upper and lower curves represent estimates of this threshold obtained by Proposition 12 and Proposition 14.

but they do provide an interesting order of magnitude for $k(l)$. In particular, we will see how the theory of large deviations and other inequalities concerning the tail of the binomial distribution can provide a sufficient condition of meaningfulness. Two main outcomes of these estimates are as follows:

- The $\log(\varepsilon)$ and $\log(N)$ dependence of this definition. In the experiments $\varepsilon = 1$, therefore allowing less than one false detection on the average. Now, the log ε-dependence guarantees that fixing a smaller or larger value for ε will change the detection result only if we change the order of magnitude of ε.
- Detection is possible with a $\sqrt{l}$ excess of alignments in a length l segment.

Let us start with an estimate of the smallest length l of a detected alignment. The first simple necessary condition we can get is a threshold on the length l. For an ε-meaningful segment, one has

$$p^l \leq \mathbb{P}[S_l \geq k(l)] \leq \frac{\varepsilon}{N^4},$$

so that

$$l \geq \frac{-4\log N + \log \varepsilon}{\log p}. \tag{5.8}$$

Let us give a numerical example: If the size of the image is $N = 512$, and if $p - 1/16$ (which corresponds to 16 possible directions), the minimal length of a 1-meaningful segment is $l_{min} = 9$.

5.3.1 Sufficient Condition of Meaningfulness

In this subsection, we will see how the theory of large deviations and the inequalities concerning the tail of the binomial distribution (see the previous chapter) can provide a sufficient condition of meaningfulness. The key point here is the result due to Hoeffding [Hoe63] (see the previous chapter: Proposition 3 and the proof in Exercise 4.3.2). They will be used to deduce a sufficient condition for a segment to be meaningful. The size N of the image and the precision p are fixed.

Proposition 12 (Sufficient condition of ε-meaningfulness) *Let S be a length l segment, containing at least k aligned points. If*

$$k \geq pl + \sqrt{\frac{4\log N - \log \varepsilon}{h(p)}} \sqrt{l},$$

where $p \mapsto h(p)$ is the function defined as in Proposition 3 by

$$h(p) = \frac{1}{1 - 2p} \log \frac{1 - p}{p} \quad if \quad 0 < p < \frac{1}{2},$$

$$h(p) = \frac{1}{2p(1 - p)} \quad if \quad \frac{1}{2} \leq p < 1,$$

then S is ε-meaningful.

Proof — Let S be a length l segment, containing at least k aligned points, where k and l are such that

$$k \geq pl + \sqrt{\frac{4 \log N - \log \varepsilon}{h(p)}} \sqrt{l}.$$

If we denote $r = k/l$, then $r \geq p$ and

$$l(r-p)^2 \geq \frac{4 \log N - \log \varepsilon}{h(p)}.$$

By Proposition 3 we deduce that

$$\mathbb{P}[S_l \geq k] \leq \exp(-l(r-p)^2 h(p)) \leq \exp(-4 \log N + \log \varepsilon) = \frac{\varepsilon}{N^4},$$

which means by definition that the segment S is ε-meaningful.   □

Corollary 1. *Let S be a length l segment, containing at least k aligned points. If*

$$k \geq pl + \sqrt{\frac{l}{2}(4 \log N - \log \varepsilon)},$$

then S is ε-meaningful.

Proof — This result is a simple consequence of Proposition 12 and of the fact that for p in $(0,1)$, $h(p) \geq 2$ (see Proposition 3 and Figure 4.1 in the previous chapter).   □

5.3.2 Asymptotics for the Meaningfulness Threshold $k(l)$

In this section ε and p are fixed. We will work on asymptotic estimates of $k(l)$ when l is "large". For this, we could use the following proposition, which is the Central Limit Theorem in the particular case of the binomial distribution (see [Fel68]).

Proposition 13 (De Moivre-Laplace limit theorem) *If α is a fixed positive number, then as l tends to $+\infty$,*

$$\mathbb{P}\left[S_l \geq pl + \alpha \sqrt{l \cdot p(1-p)}\right] \longrightarrow \frac{1}{\sqrt{2\pi}} \int_{\alpha}^{+\infty} e^{-x^2/2} \, dx.$$

Now, the problem is that if l tends to infinity, we also have to consider that N tends to infinity (indeed, since l is the length of a segment in a $N \times N$ image, $l \leq \sqrt{2}N$). And so the parameter α used in the De Moivre-Laplace theorem will depend on N. This is why a stronger version of the previous theorem, Proposition 5, is useful.

Proposition 14 (Asymptotic behavior of $k(l)$) When $N \to +\infty$ and $l \to +\infty$ in such a way that $l/(\log N)^3 \to +\infty$, then

$$k(l) = pl + \sqrt{2p(1-p) \cdot l \cdot \left(\log \frac{N^4}{\varepsilon} + O(\log \log N) \right)}.$$

Proof — We define, for $i \in \{0, 1\}$,

$$\alpha_i(l, N) = \frac{k(l) - i - pl}{\sqrt{lp(1-p)}}.$$

Corollary 1 implies that

$$k(l) \le pl + \sqrt{\frac{l}{2}(4 \log N - \log \varepsilon) + 1}$$

from which we deduce that

$$\frac{\alpha_i^6(l, N)}{l} \le C \frac{(4 \log N - \log \varepsilon)^3}{l},$$

where C is a constant. Since ε is fixed and $l/(\log N)^3 \to +\infty$, we get that $\alpha_i^6(l, N)/l \to 0$. Hence, we can apply the generalized Central Limit Theorem (Proposition 5) to obtain

$$\forall i \in \{0, 1\}, \quad \mathbb{P}\left[S_l \ge pl + \alpha_i(l, N) \sqrt{l \cdot p(1-p)} \right] \underset{l \to +\infty}{\sim} \frac{1}{\sqrt{2\pi}} \int_{\alpha_i(l,N)}^{+\infty} e^{-x^2/2} \, dx.$$

$$(5.9)$$

For $i = 0$ (resp. for $i = 1$), the left-hand term of (5.9) is smaller (resp. larger) than ε/N^4. Additionnaly, the right-hand term is equivalent to

$$\frac{1}{\sqrt{2\pi}\alpha_i(l, N)} e^{-\alpha_i^2(l,N)/2}.$$

For $i = 0$, we deduce that

$$\frac{1}{\sqrt{2\pi}} \frac{1}{\alpha_0(l, N)} e^{-\alpha_0^2(l,N)/2}(1 + o(1)) \le \frac{\varepsilon}{N^4},$$

which implies

$$O(1) + O(\log(\alpha_0(l, N))) - \frac{\alpha_0^2(l, N)}{2} + o(1) \le \log \frac{\varepsilon}{N^4}$$

and, finally,

$$\alpha_0(l, N)^2 \ge 2 \log \frac{N^4}{\varepsilon} + O(\log \log N)$$

that is

$$k(l) \geq pl + \sqrt{2p(1-p) \cdot l \cdot \left(\log \frac{N^4}{\varepsilon} + O(\log\log N) \right)}. \qquad (5.10)$$

The case $i = 1$ gives, in a similar way,

$$k(l) - 1 \leq pl + \sqrt{2p(1-p) \cdot l \cdot \left(\log \frac{N^4}{\varepsilon} + O(\log\log N) \right)}. \qquad (5.11)$$

Finally (5.10) and (5.11) yield the estimate of $k(l)$ announced in Proposition 14. $\square$

5.3.3 Lower Bound for the Meaningfulness Threshold $k(l)$

In this part, a necessary condition of ε-meaningfulness is obtained by using the comparison between the binomial and the Gaussian laws given in Slud's Theorem 1.

Proposition 15 (Necessary condition of meaningfulness) *We assume that p and N are fixed, with $0 < p \leq 1/4$ and $pN^4 > 1$. If a segment S with length l and containing k aligned points is ε-meaningful, then*

$$k \geq pl + \alpha(N)\sqrt{lp(1-p)},$$

where $\alpha(N)$ is uniquely defined by

$$\frac{1}{\sqrt{2\pi}} \int_{\alpha(N)}^{+\infty} e^{-x^2/2} \, dx = \frac{\varepsilon}{N^4}.$$

Proof — This proposition is a direct consequence of Proposition 11 (which implies that $k > pl$) and of Slud's theorem (Theorem 1). The assumption $0 < p \leq 1/4$ is not a strong condition since it is equivalent to considering that the number of possible oriented directions is larger than 4. Let us denote

$$\beta(k,l) = \frac{k - pl}{\sqrt{lp(1-p)}}.$$

Then

$$\frac{\varepsilon}{N^4} = \frac{1}{\sqrt{2\pi}} \int_{\alpha(N)}^{+\infty} e^{-\frac{x^2}{2}} \, dx \geq \mathbb{P}\left[S_l \geq k\right] \geq \frac{1}{\sqrt{2\pi}} \int_{\beta(k,l)}^{+\infty} e^{-\frac{x^2}{2}} \, dx, \qquad (5.12)$$

where the first equality defines $\alpha(N)$, the second inequality traduces that the segment S is ε-meaningful, and the third inequality follows from Slud's theorem. This implies $\beta(k,l) \geq \alpha(N)$, which is the announced inequality. $\square$

5.4 Properties of Meaningful Segments

5.4.1 Continuous Extension of the Binomial Tail

Following [Fel68], the discrete function $\mathcal{B}(l,k)$ will be extended to a continuous domain. This is done by introducing a new function $\tilde{\mathcal{B}}(l,k)$, which is called the "incomplete Beta function" in the literature.

Lemma 1 *The map*

$$\tilde{\mathcal{B}} : (l,k) \mapsto \frac{\int_0^p x^{k-1}(1-x)^{l-k}\,dx}{\int_0^1 x^{k-1}(1-x)^{l-k}\,dx} \tag{5.13}$$

is continuous on the domain $\{(l,k) \in \mathbb{R}^2, \quad 0 \le k \le l < +\infty\}$, *decreasing with respect to* k, *increasing with respect to* l, *and for all integer values of* k *and* l, *one has* $\tilde{\mathcal{B}}(l,k) = \mathcal{B}(l,k)$.

Proof — More details are available in Exercise 5.7.2 at the end of the chapter. The continuity results from classical theorems on the regularity of parameterized integrals. Notice that the continuous extension of $\tilde{\mathcal{B}}$ when $k = 0$ is $\tilde{\mathcal{B}}(l,0) = 1$. Let us prove that $\tilde{\mathcal{B}}(l,k)$ is decreasing with respect with k. The proof involves the function

$$A(l,k) = \frac{\int_0^p x^{k-1}(1-x)^{l-k}\,dx}{\int_p^1 x^{k-1}(1-x)^{l-k}\,dx}.$$

Notice that $1/\tilde{\mathcal{B}} = 1 + 1/A$. It must be proven that A decreases with respect with k. Computing

$$\frac{1}{A}\frac{\partial A}{\partial k}(l,k) = \frac{\int_0^p x^{k-1}(1-x)^{l-k} \cdot \log \frac{x}{1-x}\,dx}{\int_0^p x^{k-1}(1-x)^{l-k}\,dx} - \frac{\int_p^1 x^{k-1}(1-x)^{l-k} \cdot \log \frac{x}{1-x}\,dx}{\int_p^1 x^{k-1}(1-x)^{l-k}\,dx},$$

and applying the Mean Value Theorem implies the existence of (α, β) such that

$$0 < \alpha < p < \beta < 1 \quad \text{and} \quad \frac{1}{A}\frac{\partial A}{\partial k}(l,k) = \log \frac{\alpha}{1-\alpha} - \log \frac{\beta}{1-\beta}.$$

The right-hand term being negative, the proof is complete. The proof that $\mathcal{B}$ increases with respect with l is similar, the increasing map $x \mapsto \log \frac{x}{1-x}$ being replaced by the decreasing map $x \mapsto \log(1-x)$. Finally, the fact that $\tilde{\mathcal{B}}(l,k) = \mathcal{B}(l,k)$ for integer values of k and l is a consequence of the relation $\tilde{\mathcal{B}}(l+1,k+1) = p\tilde{\mathcal{B}}(l,k) + (1-p)\tilde{\mathcal{B}}(l,k+1)$ (see [Fel68] for example).   □

Remark: The properties of $\tilde{B}$ guarantee that $\tilde{B}$ is a good interpolate of B in the sense that the monotonicity of B in both variables l and k is extended to the continuous domain. Notice that a proof based on the same method (using that $x \mapsto \log x$ is increasing) will establish that

$$\frac{\partial \tilde{B}}{\partial l} + \frac{\partial \tilde{B}}{\partial k} \leq 0,$$

which is the natural extension of the property $B(l+1,k+1) \leq B(l,k)$ previously established in Proposition 10. The following property is a good example of the interest of the continuous extension of B. It yields a proof of Proposition 11.

Lemma 2 *If* $l \geq 1$ *and* $p \leq \frac{1}{2}$, *then* $p \leq \tilde{B}(l, p(l-1)+1) \leq \frac{1}{2}$.

Proof — Using $A(l,k)$ as in Lemma 1, one sees that it is sufficient to prove that if $k - 1 = p(l-1)$, then

$$\frac{p}{1-p} \int_p^1 x^{k-1}(1-x)^{l-k}\,dx \leq \int_0^p x^{k-1}(1-x)^{l-k}\,dx \leq \int_p^1 x^{k-1}(1-x)^{l-k}\,dx. \quad (5.14)$$

For this purpose, let us write $f(x) = x^{k-1}(1-x)^{l-k}$ and study the map

$$g(x) = \frac{f(p-x)}{f(p+x)}.$$

By a simple computation one sees that $g'(x)$ has the sign of

$$2x^2(k-1-(1-p)(l-1)) - 2p(1-p)(k-1-p(l-1))$$

and since $k - 1 = p(l-1)$ and $p \leq 1/2$, we obtain $g' \leq 0$ on $]0,p]$. Hence, $g(x) \leq g(0) = 1$ on $]0,p]$ which implies

$$\int_0^p f(x)\,dx = \int_0^p f(p-x)\,dx \leq \int_0^p f(p+x)\,dx = \int_p^{2p} f(x)\,dx \leq \int_p^1 f(x)\,dx$$

and the right-hand side of (5.14) is proven.

For the left-hand side, one can follow the same reasoning with the map

$$g(x) = \frac{f(p-x)}{f(p+\frac{1-p}{p}x)}.$$

After a similar computation, one obtains $g' \geq 0$ on $]0,p]$, so that $f(p-x) \geq f(p+\frac{1-p}{p}x)$ on $]0,p]$. Integrating this inequality yields

$$\int_0^p f(x)\,dx = \int_0^p f(p-x)\,dx \geq \int_0^p f\left(p+\frac{1-p}{p}x\right)\,dx = \frac{p}{1-p}\int_p^1 f(x)\,dx,$$

which proves the left-hand side of (5.14). $\qquad\qquad\qquad\qquad\qquad\qquad\qquad\square$

5.4.2 Density of Aligned Points

In general, it is not easy to compare $\mathcal{B}(l,k)$ and $\mathcal{B}(l',k')$ by performing simple computations on k, k', l, and l'. Assume that we have observed a meaningful segment S in a $N \times N$ image. Let l denote the length of this segment and let k denote the number of aligned points it contains. For simplicity, the segment S will be denoted by $S = (l,k)$. Assume also that we have been able take a better photograph that increases the resolution of the image. The new image has size $\lambda N \times \lambda N$, with $\lambda > 1$. Assume that the relative density of the aligned points on the segment does not change in the more resolute image. The considered segment is therefore $S_\lambda = (\lambda l, \lambda k)$. Our aim is to compare the NFAs of S and S_λ. This leads one to compare

$$N^4 \cdot \tilde{\mathcal{B}}(l,k) \quad \text{and} \quad (\lambda N)^4 \cdot \tilde{\mathcal{B}}(\lambda l, \lambda k).$$

The result given in the following theorem shows that

$$\text{NFA}(S_\lambda) < \text{NFA}(S).$$

This is a consistency check for the Helmholtz principle, since a better view must increase the detection rate and therefore decrease the NFA!

Theorem 4 *Let $S = (l,k)$ be a 1-meaningful segment of a $N \times N$ image (with $N \geq 6$). Then the function defined for $\lambda \geq 1$ by*

$$\lambda \mapsto (\lambda N)^4 \cdot \tilde{\mathcal{B}}(\lambda l, \lambda k)$$

decreases.

This theorem has the following corollary, which enables us to compare the "meaningfulness" of two segments of the same image.

Corollary 2. *Let $A = (l,k)$ and $B = (l',k')$ be two 1-meaningful segments of a $N \times N$ image (with $N \geq 6$) such that*

$$\frac{k'}{l'} \geq \frac{k}{l} \quad \text{and} \quad l' > l.$$

Then B is more meaningful than A; that is, $\text{NFA}(B) < \text{NFA}(A)$.

Proof — Indeed, we can take $\lambda = l'/l > 1$, so that $k' \geq \lambda k$. We then have by Theorem 4

$$(\lambda N)^4 \tilde{\mathcal{B}}(l',k') \leq N^4 \tilde{\mathcal{B}}(l,k),$$

and therefore $N^4 \tilde{\mathcal{B}}(l',k') < N^4 \tilde{\mathcal{B}}(l,k)$ (i.e., $\text{NFA}(B) < \text{NFA}(A)$). □

An interesting application of Corollary 2 is the concatenation of meaningful segments. Let $A = (l,k)$ and $B = (l',k')$ be two meaningful segments lying on the same line. Assume that A and B are consecutive, so that $A \cup B$ is simply a $(l+l', k+k')$ segment. Then since

$$\frac{k+k'}{l+l'} \geq \min\left(\frac{k}{l}, \frac{k'}{l'}\right),$$

we deduce, thanks to the above corollary, that

$$\mathrm{NFA}(A \cup B) < \max(\mathrm{NFA}(A), \mathrm{NFA}(B)).$$

This proves the following corollary.

Corollary 3. *The concatenation of two meaningful segments is more meaningful than the least meaningful of both.*

The next lemma is useful to prove Theorem 4.

Lemma 3 *Define for $p < r \leq 1$, $B(l,r) = \tilde{\mathcal{B}}(l, rl)$. Then*

$$\frac{1}{B} \frac{\partial B}{\partial l} < \frac{1}{l} - (g_r(r) - g_r(p)),$$

where g_r is the function defined by $x \mapsto g_r(x) = r \log x + (1 - r) \log(1 - x)$.

Proof — We first write the Beta integral in terms of the Gamma function (see, e.g., [Ana65]),

$$\int_0^1 t^{x-1}(1-t)^{y-1}\, dt = \frac{\Gamma(x)\Gamma(y)}{\Gamma(x+y)}.$$

Thanks to the definition of $\tilde{\mathcal{B}}(l, rl)$, this yields

$$B(l,r) = \frac{\Gamma(l+1)}{\Gamma(rl)\Gamma((1-r)l+1)} \int_0^p x^{rl-1}(1-x)^{(1-r)l}\, dx. \tag{5.15}$$

We now use the expansion (see [Ana65])

$$\frac{d \log \Gamma(x)}{dx} = -\gamma - \frac{1}{x} + \sum_{n=1}^{+\infty}\left(\frac{1}{n} - \frac{1}{x+n}\right), \tag{5.16}$$

where γ is Euler's constant. Using (5.15) and (5.16), we obtain

$$\frac{1}{B}\frac{\partial B}{\partial l} = -\gamma - \frac{1}{l+1} + \sum_{n=1}^{+\infty}\left(\frac{1}{n} - \frac{1}{l+1+n}\right) - r\left[-\gamma - \frac{1}{rl} + \sum_{n=1}^{+\infty}\left(\frac{1}{n} - \frac{1}{rl+n}\right)\right]$$

$$-(1-r)\left[-\gamma - \frac{1}{(1-r)l+1} + \sum_{n=1}^{+\infty}\left(\frac{1}{n} - \frac{1}{(1-r)l+1+n}\right)\right]$$

$$+ \frac{\int_0^p (r\log x + (1-r)\log(1-x)) x^{rl-1}(1-x)^{(1-r)l}\, dx}{\int_0^p x^{rl-1}(1-x)^{(1-r)l}\, dx}.$$

The function $x \mapsto r\log x + (1-r)\log(1-x)$ is increasing on $(0,r)$ and we have $p < r$, so

$$\frac{\int_0^p (r\log x + (1-r)\log(1-x))x^{rl-1}(1-x)^{(1-r)l}\,dx}{\int_0^p x^{rl-1}(1-x)^{(1-r)l}\,dx} \leq r\log p + (1-r)\log(1-p)\,.$$

Then

$$\frac{1}{B}\frac{\partial B}{\partial l} \leq \frac{1}{l} + \sum_{n=1}^{+\infty}\left(\frac{r}{rl+n} + \frac{1-r}{(1-r)l+n} - \frac{1}{l+n}\right) + r\log p + (1-r)\log(1-p)\,.$$

Now, let us consider the function

$$f : x \mapsto \frac{r}{rl+x} + \frac{1-r}{(1-r)l+x} - \frac{1}{l+x}$$

defined for all $x > 0$. Since $0 < r \leq 1$, we have $rl + x \leq l + x$ and $(1-r)l + x \leq l + x$, so that $f(x) \geq 0$ and

$$f'(x) = -\frac{r}{(rl+x)^2} - \frac{1-r}{((1-r)l+x)^2} + \frac{1}{(l+x)^2} \leq 0.$$

We deduce that for N integer larger than 1,

$$\sum_{n=1}^{N} f(n) \leq \int_0^N f(x)\,dx\,.$$

A simple integration gives

$$\int_0^N f(x)\,dx = r\log\left(1+\frac{rl}{N}\right) + (1-r)\log\left(1+\frac{(1-r)l}{N}\right)$$
$$- \log\left(1+\frac{l}{N}\right) - r\log r - (1-r)\log(1-r).$$

Finally,

$$\sum_{n=1}^{+\infty}\left(\frac{r}{rl+n} + \frac{1-r}{(1-r)l+n} - \frac{1}{l+n}\right) \leq -r\log r - (1-r)\log(1-r),$$

which yields

$$\frac{1}{B}\frac{\partial B}{\partial l} \leq \frac{1}{l} - r\log r - (1-r)\log(1-r) + r\log p + (1-r)\log(1-p)$$
$$= \frac{1}{l} - g_r(r) + g_r(p).$$

$\square$

Proof of Theorem 4 — Let us define $r = k/l$. Since S is 1-meaningful, we have $r > p$ and also (see Lemma 4 in Exercise 5.7.3)

$$g_r(r) - g_r(p) \geq \frac{3 \log N}{l}.$$

Let f be the function defined for $\lambda \geq 1$ by $f(\lambda) = (\lambda N)^4 \tilde{\mathcal{B}}(\lambda l, \lambda k) = (\lambda N)^4 B(\lambda l, r)$. If we compute the derivative of f and use Lemma 3, we get

$$\frac{\partial \log f}{\partial \lambda} = \frac{4}{\lambda} + l \frac{\partial \log B}{\partial l}(\lambda l, r)$$

$$< \frac{4}{\lambda} + l \left(\frac{1}{\lambda l} - g_r(r) + g_r(p) \right)$$

$$< \frac{5}{\lambda} - 3 \log N,$$

which is negative thanks to the hypothesis $N \geq 6$. $\qquad\qquad\square$

Remark: For the approximation of $\tilde{\mathcal{B}}(l, k)$ given by the Gaussian law

$$G(l,k) = \frac{1}{\sqrt{2\pi}} \int_{\alpha(l,k)}^{+\infty} e^{-\frac{x^2}{2}} dx \qquad \text{where} \qquad \alpha(l,k) = \left(\frac{k}{l} - p \right) \sqrt{\frac{l}{p(1-p)}},$$

we immediately have the result that $G(l',k') < G(l,k)$ when $k'/l' \geq k/l > p$ and $l' > l$.

5.5 About the Precision p

In this subsection the problem of the choice of the precision p is addressed. It will be shown that it is useless to make p too small; this would yield no better detection rates.

Consider a segment S of length l. We can assume that the direction of the segment is $\theta = 0$. Suppose that among the l points one observes k aligned points with given precision p (i.e., k points having their direction in $[-p\pi, +p\pi]$). Now, what happens if we change the precision p into $p' < p$? Knowing that there are k points with direction in $[-p\pi, +p\pi]$, we can assume (by the Helmholtz principle) that the average number of points having their direction in $[-p'\pi, p'\pi]$ is $k' = \frac{p'}{p}k$. Our aim is to compare

$$\tilde{\mathcal{B}}(l, k, p) \qquad \text{and} \qquad \tilde{\mathcal{B}}(l, k', p'),$$

where $\tilde{\mathcal{B}}(l, k, p)$ denotes what we wrote anteriorly $\tilde{\mathcal{B}}(l, k)$, when p was fixed. Since we are interested in meaningful segments, we will only consider the case

$$\lambda = \frac{k}{l \cdot p} = \frac{k'}{l \cdot p'} > 1.$$

One is led to study the function $p \longmapsto \tilde{B}(l, \lambda l p, p)$ and check that it is decreasing. This question is open. Fortunately, an analogous statement using the large deviations estimate of the binomial tail (Proposition 6) is accessible.

Proposition 16 *Consider the large deviations estimate of Proposition 6, given by*

$$L(l,k,p) = -l\left(\frac{k}{l}\log\frac{k}{lp} + \left(1 - \frac{k}{l}\right)\log\frac{1 - \frac{k}{l}}{1 - p}\right);$$

then for any $\lambda > 1$, the map $p \mapsto L(l, \lambda l p, p)$ is decreasing with respect to p.

Proof — One can easily prove that the function

$$p \longmapsto \lambda p \log \lambda + (1 - \lambda p) \log \frac{1 - \lambda p}{1 - p}$$

increases (for $\lambda > 1$). Consequently,

$$p \longmapsto L(l, \lambda l p, p) = -l\left(\frac{\lambda l p}{l}\log\frac{\lambda l p}{lp} + \left(1 - \frac{\lambda l p}{l}\right)\log\frac{1 - \frac{\lambda l p}{l}}{1 - p}\right)$$

decreases.   □

This result is experimentally checked in Figure 5.6, where we compute the maximal meaningful alignments of an image first for the usual precision $p = 1/16$ and then for $p = 1/32$. Almost all of the alignments detected at precision $1/32$ are already detected at precision $1/16$. The previous argument shows that we must always take the precision as coarse as possible. In fact, there is another inconvenience in choosing very fine precisions such as $p = 1/64$. Spurious meaningful alignments are then detected. They are due to the quantization of gray levels of the image, which creates a quantization of the orientations. This problem is addressed in [DLMM02].

5.6 Bibliographic Notes

The notion of ε-meaningful segments has to be related to the classical "α-significance" in statistics, where α is simply $w(l, \varepsilon, N) = \frac{\varepsilon}{N^4}$. The reason leading us to a slightly different terminology is that we are not in a position to assume that segments detected as ε-meaningful are independent. Indeed, if a segment is very meaningful, it will be contained in many larger segments that also are ε-meaningful. Thus, it is convenient to compare the number of detected segments to the expectation of this number. This is not exactly the same situation as in failure detection, where the failures are somehow disjointed and rare events. This new definition overcomes a difficulty raised by Stewart [Ste95] in his seminal work on the 'MINPRAN" method. The method was presented as a new paradigm and applied to the 3-D alignment problem. It is worth describing the method in some detail and explaining what is

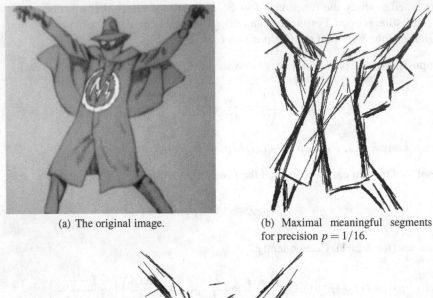

(a) The original image.   (b) Maximal meaningful segments
                                       for precision $p = 1/16$.

(c) Maximal meaningful segments
for precision $p = 1/32$.

Fig. 5.6 Increasing the precision: $p = 1/16$ and $p = 1/32$. Almost all of the alignments detected at precision $1/32$ are already detected at precision $1/16$. In this figure, only maximal meaningful segments are displayed. They will be defined in Chapter 6 and they are in some sense the "best representatives" of the set of all meaningful segments.

being added to it. To start with, a hypothesis of 3-D alignment on a plane P is generated. Let us call r the distance to the plane and consider the event "at least k points among the N randomly fall within the range $P \pm r$" (i.e. at a distance less than r from P). The probability of the event is, calling the maximal distance to the plane z_0,

$$\mathcal{B}\left(N, k, \frac{r}{z_0}\right) = \sum_{i=k}^{N} \binom{N}{i} \left(\frac{r}{z_0}\right)^i \left(1 - \frac{r}{z_0}\right)^{N-i}.$$

Then for a given plane P, Stewart computes the minimal probability of alignment over all r's; that is,

$$H(P,N) = \min_r \mathcal{B}\left(N, k_{P,r}, \frac{r}{z_0}\right).$$

Then the number of hypothesized planes S is fixed. We quote:

MINPRAN accepts the best fit from S samples as correct if

$$\min_{1 \le j \le S} H(P_j, N) < H_0,$$

where H_0 is a threshold based on the probability P_0 that the best fit to N uniformly distributed outliers is less than H_0. Intuitively, P_0 is the probability MINPRAN will hallucinate (sic) a fit where there is none. Thus, for a user defined value of P_0 (for example $P_0 = 0.05$) we establish our threshold value H_0 (...) To make the analysis feasible, we assume the S fits and their residuals are independent. Strictly speaking, this assumption is not correct since the point set is the same for all fits. It is reasonable for relatively small values of S (...). Doing this requires several parameters to be specified by the user. These parameters (...) are the estimated maximum fraction of true outliers (...), the minimum number of points to be allowed in a fit, and the estimated maximum number of correct fits.

Stewart's method starts exactly as the method introduced here. Stewart addresses but does not solve the two problems overcome in the present chapter. One is the generation of the set of samples, which in Stewart's method leads to the involvement of at least *three* user's parameters. The second one is the severe restriction about the *independence of samples*. Both difficulties were simultaneously solved in this chapter by introducing the number of samples as an implicit parameter of the method (computed from the image size and Shannon's principles) and by replacing in all calculations the "probability of hallucinating a wrong event" by the "expectation of the number of such hallucinations" or false alarm rate. Here is how this chapter's method would find "alignments" on planes. It would first sample the set of all planes, according to the accuracy of the image. Then the NFA of a given plane P would be $\text{NFA}(P) = NH(P,N)$, where N is the number of hypothesized planes. A plane P is ε-meaningful if $\text{NFA}(P) < \varepsilon$ and the whole former theory applies. The parameter ε plays the role of P_0. The term ε-*meaningful* is related to the classical *p-significance* in statistics; ε can also be seen as the equivalent of the risk level in statistical hypothesis testing. We refer to [DMM00] and [DMM03c] as the main source of this chapter.

As we mentioned previously, it is possible to define a number of false alarm in a general way, as done recently in [GM06]. According to their formulation, a function $F(i,x)$ is a NFA associated to the random variables $(X_i)_{i \in I}$ as soon as one has

$$\forall \varepsilon > 0, \quad \mathbb{E}\left[\left|\{i, F(i, X_i) \le \varepsilon\}\right|\right] \le \varepsilon. \tag{5.17}$$

Then, thanks to the subuniform distribution of p-values, one can see easily that the function

$$F(i, x_i) = n_i \cdot \mathbb{P}\left[X_i \ge x_i\right]$$

is a NFA as soon as

$$\sum_{i \in I} \frac{1}{n_i} \leq 1, \tag{5.18}$$

which yields a single proof for all of results like Proposition 9. Moreover, if (5.18) is an equality and the X_i's admit density functions, then (5.17) is also an equality.

The detection of alignments is made in image analysis by using the Hough Transform (see [Mai85]). More generally, the detection of globally salient structures has been addressed by Sha'Ashua and Ullman (see [SU88]). Let us also mention the Extension Field of Guy and Medioni (see [GM96]) and the Parent and Zucker curve detector (see [PZ89]). These methods have, however, the same drawback as most image analysis methods. They *a priori* suppose that what they want to find (lines, circles, curves, etc.) is in the image. So they may find too many or too little such structures in the image and therefore do not yield an *existence proof* for the found structures. Let us describe in more detail the Hough Transform. Assume that the image under analysis is made of dots, that may create aligned patterns or not. The result of the Hough Transform is a map associating with each line its number of dots. The peaks of the Hough Transform indicate the lines that have more dots. Which peaks are significant? Clearly, a threshold must be used and we are led back to the problems addressed here. The analysis of the Hough Transform performed by Kiryati, Eldar, and Bruckstein [KEB91] and by Shaked, Yaron, and Kiryati [SYK96] is very close to the analysis in this chapter. These authors prove by large deviations estimates that lines in an image detected by the Hough Transform could be detected as well in an undersampled image without increasing significantly the false alarm rate. They view this method as an accelerator tool, whereas it is developed it here as a detection tool. The essentials of their analysis are contained in Section 5.4.2. They can be summarized as: *the closer one sees, the better one detects.*

A clear limitation of the alignment detection method developed in this chapter is the use of binary variables for point alignments, depending on the precision parameter p. The event that k among l points are aligned at precision p does not take advantage of the fact that many of the k points may be much more precisely aligned than the bound p. Thus, it seems useful to define a soft threshold function $f(\theta_i)$, where θ_i is the angle between the gradient at a point i of the considered segment and the normal to the segment. A soft threshold is an even function $[-\pi, \pi] \to [0, 1]$ such that $f(0) = 0$, $f(\pm\pi) = 1$, and f is increasing on $[0, \pi]$. An ε-meaningful alignment is a segment such that $N^4 \mathbb{P}(\sum_{i=1}^{l} f(\theta_i) \leq \eta) \leq \varepsilon$, where the θ_i are the random orientation variables, uniformly distributed in the a-contrario model and η is the observed value on the considered segment of $\sum_{i=1}^{l} f(\theta_i)$. In [Igu06], this idea is developed and leads to the definition of a "continuous" NFA. An accurate enough estimate of the law of $\sum_{i=1}^{l} f(\theta_i)$ is obtained by a generalization of Hoeffding's inequality.

In [AFI+06], the authors have applied 3-D alignment detection in the spirit of the present chapter to recover urban models from noisy 3-D data. They apply a region merging technique. The merging criterion involves the NFA of the hypothesis: "All points in the region are aligned with a plane."

5.7 Exercises

5.7.1 Elementary Properties of the Number of False Alarms

Set $\mathcal{B}(l,k_0,p) = \sum_{k=k_0}^{l} \binom{l}{k} p^k (1-p)^{l-k} = \mathbb{P}[S_l \geq k]$ and $\mathrm{NFA}(l,k) = N^4 \mathcal{B}(l,k,p)$. This last number is called "number of false alarms for an alignment of k points among l with precision p in a N^2 pixels image". Prove and interpret in this framework the following properties:

1. $\mathrm{NFA}(l,0) = N^4$.
2. $\mathrm{NFA}(l,l) = N^4 p^l$. Compute the minimal size of a 1-meaningful segment having all points aligned when $N = 512$, $p = \frac{1}{16}$.
3. $\mathrm{NFA}(l,k+1) < \mathrm{NFA}(l,k)$.
4. $\mathrm{NFA}(l,k) < \mathrm{NFA}(l+1,k)$.
5. $\mathrm{NFA}(l+1,k+1) < \mathrm{NFA}(l,k)$.

5.7.2 A Continuous Extension of the Binomial Law

The parameter p being fixed, we set $\mathcal{B}(l,k) = \mathbb{P}[S_l \geq k]$ and

$$\tilde{\mathcal{B}}(l,k) = \frac{\int_0^p x^{k-1}(1-x)^{l-k} dx}{\int_0^1 x^{k-1}(1-x)^{l-k} dx}. \tag{5.19}$$

1) Prove that $\tilde{\mathcal{B}}$ is continuous in the domain $\{(l,k) \in \mathbb{R}^2, \ 0 \leq k \leq l < +\infty\}$. (You will have to define $\tilde{\mathcal{B}}(l,0)$ as a limit when l tends to zero).

2) Check that $\tilde{\mathcal{B}}$ is decreasing with respect to k. *Hint:* Set

$$A(l,k) = \frac{\int_0^p x^{k-1}(1-x)^{l-k} dx}{\int_p^1 x^{k-1}(1-x)^{l-k} dx}$$

and notice that $1/\tilde{\mathcal{B}} = 1 + 1/A$. Show that

$$\frac{1}{A}\frac{\partial A}{\partial k}(k,l) = \frac{\int_0^p x^{k-1}(1-x)^{l-k} \cdot \log \frac{x}{1-x} dx}{\int_0^p x^{k-1}(1-x)^{l-k} dx} - \frac{\int_p^1 x^{k-1}(1-x)^{l-k} \cdot \log \frac{x}{1-x} dx}{\int_p^1 x^{k-1}(1-x)^{l-k} dx}.$$

By applying the Mean Value Theorem for integrals, deduce that there are (α, β) such that

$$0 < \alpha < p < \beta < 1 \qquad \text{and} \qquad \frac{1}{A}\frac{\partial A}{\partial k}(k,l) = \log \frac{\alpha}{1-\alpha} - \log \frac{\beta}{1-\beta}.$$

3) Prove that $\tilde{\mathcal{B}}$ is increasing with respect to l. *Hint:* Proceed like in question 2 and use $x \mapsto \log(1-x)$ instead of $x \mapsto \log \frac{x}{1-x}$.

4) Prove that for all integer values of l and k, $\tilde{\mathcal{B}}(l,k) = \mathcal{B}(l,k)$. *Hint:* Set $\tilde{\mathcal{B}}(l,k) = \frac{A(l,k)}{D(l,k)}$ and prove by integration by parts that

$$A(l,k-1) = \frac{l-k+1}{k-1}A(l,k) + \frac{p^{k-1}(1-p)^{l-k+1}}{k-1},$$

$$D(l,k) = \frac{1}{k\binom{l}{k}} = \frac{k-1}{l-k+1}D(l,k-1).$$

Deduce that

$$\tilde{\mathcal{B}}(l,k-1) = \tilde{\mathcal{B}}(l,k) + \binom{l}{k-1}p^{k-1}(1-p)^{l-k+1}.$$

Find a formula for $\tilde{\mathcal{B}}(l,l)$ and deduce recursively for $k = l, l-1, ...,$ that $\tilde{\mathcal{B}}(l,k) = \mathcal{B}(l,k)$ for $l,k \in \mathbb{N}$.

5.7.3 A Necessary Condition of Meaningfulness

Let $0 < r < 1$ be a real number and let g_r be the function defined on $(0,1)$ by

$$g_r(x) = r\log x + (1-r)\log(1-x).$$

1) Prove that g_r is concave and has its maximum at point $x = r$. Moreover, prove that if $0 < p \leq r$, then

$$2(r-p)^2 \leq g_r(r) - g_r(p) \leq \frac{(r-p)^2}{p(1-p)}.$$

The aim of this exercise is to prove the following result.

Lemma 4 *If $N \geq 5$ and if S is an ε-meaningful segment with length l and with $1 \leq k \leq l$ aligned points according to precision p, then if we denote $r = k/l$, we have*

$$g_r(r) - g_r(p) > \frac{3\log N - \log \varepsilon}{l}.$$

2) Prove that

$$\binom{l}{k} p^k (1-p)^{l-k} \leq \mathcal{B}(l,k,p) \leq \frac{\varepsilon}{N^4}.$$

3) We assume that $k \leq l-1$. Use the following inequalities (refinement of Stirling's formula, see [Fel68] for example), valid for n integer larger than 1:

$$n^n e^{-n} \sqrt{2\pi n} e^{1/(12n+1)} \leq n! \leq n^n e^{-n} \sqrt{2\pi n} e^{1/12n},$$

to prove that

$$\binom{l}{k} p^k (1-p)^{l-k} \geq \frac{2}{\sqrt{2\pi l}} e^{-1/6} e^{l(g_r(p)-g_r(r))}.$$

4) Prove the announced result in the case $k \leq l-1$.
Hint: Since the size of the considered image is $N \times N$ and l is a length of a segment of the image, we have $l \leq \sqrt{2}N$.

5) Prove directly the result when $k = l$.

6) A sufficient condition of meaningfulness.
Let S be a segment of length l containing k aligned points. Assume that $k \geq pl$ and that

$$g_r(r) - g_r(p) > \frac{4\log N - \log \varepsilon}{l}.$$

Then, using Hoeffding inequality, prove that S is ε-meaningful.

Chapter 6
Maximal Meaningfulness and the Exclusion Principle

6.1 Introduction

Alignments have been defined as segments with enough aligned points. This definition leads to a plethoric detection. Indeed, if a segment S has been detected and had a lot of aligned points, larger segments containing S also will be counted as alignments. If, in addition, the image is blurry, as it must be by the Shannon-Nyquist principle, the alignments seen in an image correspond roughly to rectangles of aligned points with 2 pixels width. Thus, the presence of a very meaningful segment will actually lead to the detection of many longer, shorter, and slanted segments, a whole bundle of meaningful segments where in fact we would be interested in only one: the best representative of the alignment. In such a case, our perception uses an economy principle and retains only the "best" alignment: the one that stands for all the other detectable ones. This question is illustrated in Figures 6.1 and 6.2, displaying all meaningful segments of a digital image. They form bundles of slanted and excessively elongated segments. The digital image in Figure 6.1 was first drawn with a ruler and pencil on a standard A4 white sheet of paper and then scanned into a 478×598 digital image. The scanner's apparent blurring kernel is about 2 pixels wide and some aliasing is also perceptible, making the strokes somewhat dashed. Two pairs of pencil strokes are aligned on purpose. In the middle image of this experiment, we display all ε-meaningful segments for $\varepsilon = 10^{-3}$. For $\varepsilon = 1$, we would have some still longer and more slanted segments. We took a value of ε smaller than 1 just to show that finding the best alignments is not solved by just decreasing ε. Three phenomena occur that are very apparent in this simple example and which are perceptible in all further experiments:

- Too long meaningful alignments. We commented on this above. Clearly, the pencil strokes boundaries are very meaningful, thus generating larger meaningful segments that contain them.
- Multiplicity of detected segments. On both sides of the strokes we find several parallel segments (reminder: the orientation of lines is modulo 2π). These parallel segments are due to the blurring effect of the scanner's optical convolution.

Fig. 6.1 Left: the original pencil strokes image. Middle: ε-meaningful alignments with $\varepsilon = 10^{-3}$. Right: maximal meaningful alignments. Maximal meaningfulness selects the right alignments on each line, but does not completely eliminate the plurality of detections for each single perceptual alignment. This will be solved by a more general version of the exclusion principle.

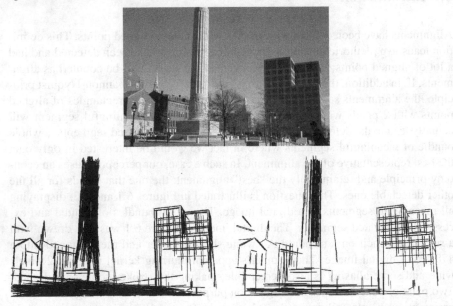

Fig. 6.2 Parallelism against alignment. Top: original Brown image. Bottom-left: maximal meaningful alignments as defined in Section 6.3. Here, since many parallel alignments are present, secondary parasite slanted alignments are also found. Bottom-right: alignments finally obtained by the exclusion principle, explained in Section 6.2, which eliminates the remaining spurious alignments. The experiment proves that maximal meaningfulness by itself cannot remove the slanted alignment detections that arise from bundles of parallel alignments. The more general exclusion principle is therefore necessary.

Classical edge detection theory would typically select the best one in terms of contrast.

– Lack of accuracy of the detected directions. We do not check whether the directions along a meaningful segment are distributed on both sides of the line's direction. Thus, it is to be expected that lines that are actually slanted with respect to the edge's "true" direction are detected. Typically, a blurry edge will generate several parallel and more or less slanted alignments.

On the right image of Figure 6.1 are displayed the maximal meaningful segments. These are selected from the meaningful segments in the middle by an economy principle. Now, perception must obey an economy principle, which we call the *exclusion principle*. According to this principle "two alignments cannot overlap".

In order to explain the generality of the principle, we will proceed in two directions. In the one way, specific to alignments, we will define "maximal meaningful segments" as segments that are local maxima of the NFA for inclusion – in other words the most meaningful segments in a list of segments ordered by inclusion. We will prove that the maximal meaningful segments indeed satisfy an *exclusion principle*: they never overlap. This solves completely and elegantly the problem for segments contained in a single line, but not for slanted segments sharing some of the same pixels. So we need a more general version of the *exclusion principle* than just maximality. We will state in Section 6.2 such a general principle and give an algorithm. The full problem and its solution in the case of alignments is illustrated in Figure 6.2, where we present first only maximal meaningful segments on each line. Clearly, overlapping slanted alignments are left. Those are easily removed by application of the exclusion principle.

Taking the easiest first, we shall define the exclusion principle and its algorithm in Section 6.2. Section 6.3 is devoted to the treatment of maximal meaningfulness of alignments, which turns out to be mathematically tricky and which leaves a mathematical conjecture open. We have added an experimental section to be able to show several experiments validating the relatively sophisticated approach (Section 6.4).

6.2 The Exclusion Principle

The general exclusion principle we will state here is applicable in a very wide set of situations.

6.2.1 Definition

Principle 4 (Exclusion principle) *Let A and B be groups obtained by the same gestalt law. Then no point x is allowed to be taken into account for both A and B when computing their NFA. In other words, each point must either "vote" for A or for B.*

The exclusion principle defines directly an algorithm.

Algorithm 4 (Computing disjoint groups with small NFA by exclusion principle)
Consider any gestalt grouping quality and let $O_1, \ldots O_n$ be groups of pixels of the image that are meaningful with respect to this quality. Such groups can overlap. The algorithm defines from them a new set of exclusive groups with smallest NFA.

– *For each $x \in \bigcup_i O_i$, fix an index $i(x)$ such that $\mathrm{NFA}(O_{i(x)})$ is minimal among all meaningful objects containing x.*
– *Define for every $k = 1, \ldots, n$, $\tilde{O}_k = \{x \in O_k, i(x) = k\}$. All $\tilde{O}_k$ are then pairwise disjoint.*
– *Compute $\mathrm{NFA}(\tilde{O}_k)$ (which is larger than $\mathrm{NFA}(O_k)$) for every k and keep only the meaningful ones.*

There is still a simpler algorithm ensuring the exclusion principle.

Algorithm 5 (Exclusion principle (EP-bis)) *In the same situation as for Algorithm 4:*

– *Pick $\tilde{O}_1 = O_{k_1}$ where $\mathrm{NFA}(O_{k_1})$ achieves the minimal NFA among the O_k's.*
– *Set for every k, $O_k^1 = O_k \setminus \tilde{O}_1$ (so that $O_{k_1}^1$ is no longer in the list.)*
– *In the same way, pick $\tilde{O}_2 = O_{k_2}^1$ where $\mathrm{NFA}(O_{k_2}^1)$ achieves the minimal NFA among the O_k^1's. If $\tilde{O}_2$ is not meaningful, stop. Otherwise, set, for every k, $O_k^2 = O_k^1 \setminus \tilde{O}_2$.*
– *Iterate until the list of objects is empty or none of them is meaningful. Keep $\tilde{O}_1, \tilde{O}_2, \ldots$ as the final groups according to the considered quality.*

In a nutshell, this algorithm takes out iteratively the most meaningful groups. Both algorithms 4 and 5 achieve the exclusion principle and we will not decide between them. The second one may seem more intuitive but the algorithm 4 yields good results for all gestalts considered in this book. It can be applied for all gestalts treated in these lectures and we will in particular apply it to vanishing points in Chapter 8.

6.2.2 Application of the Exclusion Principle to Alignments

As we saw in the introduction, one problem with the detection of meaningful alignments is that we may obtain several candidates for each perceptual segment. This is because correctly sampled images are slightly blurred, which means that edges are a little bit thicker than 1 or 2 pixels. Hence, most thin segments contained in the actual thick segment are meaningful. Among all of these segments, we are interested in selecting a single one, namely the one that best estimates its position and orientation. The exclusion principle applies to this problem.

Algorithm 6 (Exclusion principle for thick segments)

– *Compute all meaningful segments.*
– *Assign each pixel x to a segment A achieving* $\min \text{NFA}(B)$ *among all meaningful segments B lying at a distance less than r from x.*
– *Compute* $\text{NFA}(A)$ *for each meaningful segment A. Instead of the number k of observed aligned points in A, use the number* $k' \le k$ *of aligned points in A that in addition have been assigned to the segment. Therefore,* $\text{NFA}(A)$ *increases. If* $\text{NFA}(A)$ *is still smaller than* ε, *the segment is called a* maximal-EP *meaningful segment.*

In a more accurate variant, this algorithm start with the maximal meaningful segments as defined in the Section 6.3, Definition 8. Figure 6.2 illustrates the thick alignment problem and its solution by exclusion principle. In the middle, we see all detected alignments in the Brown image on the left. Clearly, those alignments make sense, but many of them are slanted. Straight edges are indeed blurry and therefore constitute a rectangular region where all points have roughly the same direction. Thus, the straight alignments are mixed up with slanted alignments, which still respect the precision bound p. We can interpret the situation as a conflict between alignment and parallelism, as already illustrated in Figure 2.27. In the right-hand image of Figure 6.2, the spurious slanted alignments are eliminated by Algorithm 6. Figures 6.3 and 6.4 are other excellent examples of the result of Algorithm 6, where all spurious segments are eliminated and only the perceptually correct ones are kept.

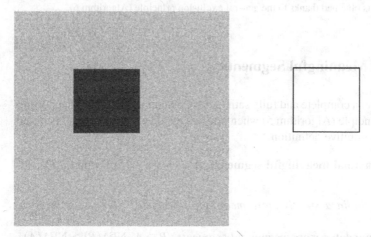

Fig. 6.3 Exclusion Principle (Algorithm 6) applied to the noisy square image. Only four segments are found corresponding to the four sides of the square. Compare this to Figure 5.4 in the previous chapter where all meaningful and all maximal meaningful segments (Algorithm 7) were computed.

Fig. 6.4 Part of original Uccello's painting image. This small image has size 216 × 162 pixels. We show respectively: all meaningful segments, maximal meaningful segments (Algorithm 7), and meaningful segments obtained thanks to the general exclusion principle (Algorithm 6).

6.3 Maximal Meaningful Segments

It is possible to get a complete and fully satisfactory mathematical theory supporting the exclusion principle (Algorithm 5) when applied inside a straight line. It is based on the following intuitive definition.

Definition 8 (Maximal meaningful segment). *A segment A is* maximal *if the following hold*

1. *It does not contain a strictly more meaningful segment:* $\forall B \subset A$, NFA$(B) \geq$ NFA(A).
2. *It is not contained in a more meaningful segment:* $\forall B \supset A$, NFA$(B) >$ NFA(A).

We say that a segment is maximal meaningful *if it is both maximal and meaningful.*

Figure 6.5 gives all maximal meaningful segments detected in a road perspective image. We are not expecting that all detected alignments correspond to perceptual straight contours. In this example, some of the maximal meaningful alignments are simply due to the slanted view, which squeezes all objects in the horizon direction.

(a) The original image.

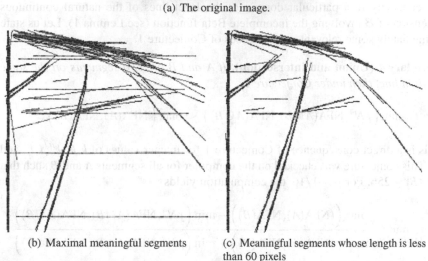

(b) Maximal meaningful segments

(c) Meaningful segments whose length is less than 60 pixels

Fig. 6.5 (a) A road (courtesy of INRETS), size 256×256; (b) all maximal meaningful segments; (c) all meaningful segments with length less than 60. The detected horizontal lines in (b) correspond to horizon lines (i.e., lines parallel to the horizon). These detections are due to a perspective effect. Indeed, all visual objects on the road (shadows, spots, etc.) are seen in a very slanted view. Thus, their contours are mostly parallel to the horizon and generate "perspective alignments". The darker dots on each gray segment indicate the aligned points.

Proposition 17 (Properties of maximal segments) *Let A be a maximal segment, then the following hold.*

1. *The two endpoints of A have their direction aligned with the direction of A.*
2. *The two points next to A (one on each side) do not have their direction aligned with the direction of A.*

Both properties follow from Proposition 10.

6.3.1 A Conjecture About Maximality

We now study the structure of maximal segments and give some evidence that two distinct maximal segments on a same straight line have no common point.

Conjecture 1 (min/max inequality for the Binomial tail) *If* $(l, l', l'') \in [1, +\infty)^3$ *and* $(k, k', k'') \in [0, l] \times [0, l'] \times [0, l'']$, *then*

$$\min \left(p, \mathcal{B}(l, k), \mathcal{B}(l + l' + l'', k + k' + k'') \right)$$
$$< \max \left(\mathcal{B}(l + l', k + k'), \mathcal{B}(l + l'', k + k'') \right). \tag{6.1}$$

This conjecture can be deduced from a stronger (but simpler) conjecture about the concavity in a particular domain of the level lines of the natural continuous extension of $\mathcal{B}$ involving the incomplete Beta function (see Lemma 1). Let us state immediately some relevant consequences of Conjecture 1.

Corollary 4 (Union and intersection). *If A and B are two segments on the same straight line, then under Conjecture 1,*

$$\min \left(pN^4, \mathrm{NFA}(A \cap B), \mathrm{NFA}(A \cup B) \right) < \max \left(\mathrm{NFA}(A), \mathrm{NFA}(B) \right).$$

This is a direct consequence of Conjecture 1 for integer values of k, k', k'', l, l', and l''. This conjecture was checked on the computer for all segments A and B such that $|A \cup B| \leq 256$. For $p = 1/16$, the computation yields

$$\min_{|A \cup B| \leq 256} \frac{\max \left((\mathrm{NFA}(A), \mathrm{NFA}(B)) \right) - \min \left(pN^4, \mathrm{NFA}(A \cap B), \mathrm{NFA}(A \cup B) \right)}{\max \left((\mathrm{NFA}(A), \mathrm{NFA}(B)) \right) + \min \left(pN^4, \mathrm{NFA}(A \cap B), \mathrm{NFA}(A \cup B) \right)}$$
$$\simeq 0.000754.$$

The minimum is independent of N and attained for $A = (23, 243), B = (23, 243)$, and $A \cap B = (22, 230)$. As usual, the couple (l, k) attached to each segment represents the number of aligned points (k) and the segment length (l).

Theorem 5 (Maximal segments are disjoint under Conjecture 1) *Suppose that Conjecture 1 is true. Then any two maximal segments lying on the same straight line have no intersection.*

Proof — Suppose that one can find two maximal segments $(l + l', k + k')$ and $(l + l'', k + k'')$ that have a nonempty intersection (l, k) Then according to Conjecture 1,

$$\min \left(p, \mathcal{B}(l, k), \mathcal{B}(l + l' + l'', k + k' + k'') \right) < \max \left(\mathcal{B}(l + l', k + k'), \mathcal{B}(l + l'', k + k'') \right).$$

If the left-hand term is equal to p, we get a contradiction since one of $(l + l', k + k')$ or $(l + l'', k + k'')$ is strictly less meaningful than the segment $(1, 1)$ it contains.

If not, we have another contradiction because one of $(l+l',k+k')$ or $(l+l'',k+k'')$ is strictly less meaningful than one of (l,k) or $(l+l'+l'',k+k'+k'')$. $\quad\square$

Remark: The numerical check of Conjecture 1 ensures that for $p = 1/16$, two maximal meaningful segments with total length smaller than 256 are disjoint. This is enough for most practical applications. Theorem 5 yields an easy algorithm for computing all meaningful alignments in a line.

Algorithm 7 (Exclusion principle for alignments in a line)

1. *Establish a candidate list of all intervals I on the line that start by an aligned point preceded by a nonaligned one and end up with an aligned point followed by a nonaligned point.*
2. *Consider in turn all pairs (I, J) where I and J belong to the list of candidates and satisfy $I \subset J$. If J is more meaningful than I or equally meaningful, remove I from the list. Iterate until no pair is left.*

Corollary 5. *Algorithm 7 computes all of the maximal meaningful segments of the line and they are disjoint.*

Proof — This is a direct application of the definition of maximal meaningful intervals and of Theorem 5. If two of the remaining intervals met, this would mean that the conjecture is wrong. The fact that only intervals with an aligned point followed by an nonaligned one can be candidates is a direct application of Proposition 10, items 4 and 5. $\quad\square$

6.3.2 A Simpler Conjecture

The rest of the section is dedicated to partial results about Conjecture 1. In this subsection, we state a simple geometric property entailing Conjecture 1.

Definition 9 (Curvature). *Let $f(x,y)$ be a real-valued function defined on some open set of $\mathbb{R}^2$. At each point where f is C^2, the curvature of f is defined by*

$$\text{curv}(f) = \frac{f_{xx}f_y^2 - 2f_{xy}f_xf_y + f_{yy}f_x^2}{(f_x^2 + f_y^2)^{\frac{3}{2}}}. \tag{6.2}$$

Note that the curvature of f at point (x_0, y_0) is nothing but the local curvature of the level line $\{(x,y),\ f(x,y) = f(x_0,y_0)\}$ at this point.

Conjecture 2 (curvature of the extended Binomial tail) *The map $(l,k) \mapsto \tilde{B}(l,k)$ defined in Lemma 1 has negative curvature on the domain*

$$D_p = \{(l,k) \in \mathbb{R}^2,\ p(l-1)+1 \leq k \leq l\}.$$

It is equivalent to say that the level curves $l \mapsto k(l,\lambda)$ of $\tilde{\mathcal{B}}$ defined by $\tilde{\mathcal{B}}(l,k(l,\lambda)) = \lambda$ are concave; that is, they satisfy

$$\forall (l_0,k_0) \in D_p, \qquad \frac{\partial^2 k}{\partial l^2}(l_0, \tilde{\mathcal{B}}(l_0,k_0)) < 0.$$

Remark: All numerical computations that we have realized so far for the function $\tilde{\mathcal{B}}(l,k)$ have been in agreement with Conjecture 2 . Concerning theoretical results, we will see in the next section that this conjecture is asymptotically true. For now, the following results show that Conjecture 2 is satisfied for the Gaussian approximation of the binomial tail (correct for small deviations i.e., $k \simeq pl + C\sqrt{l}$) and also for the large deviations estimate given by Proposition 6.

Proposition 18 *The approximation of $\mathcal{B}(l,k)$ given by the Gaussian law*

$$G(l,k) = \frac{1}{\sqrt{2\pi}} \int_{\alpha(l,k)}^{+\infty} e^{-\frac{x^2}{2}} dx \qquad where \qquad \alpha(l,k) = \frac{k-pl}{\sqrt{lp(1-p)}}$$

has negative curvature on the domain D_p.

Proof — The level lines $G(l,k) = \lambda$ of $G(l,k)$ can be written under the form

$$k(l,\lambda) = pl + f(\lambda)\sqrt{l},$$

with $f > 0$ on the domain $\{k > pl\}$. Hence, we have

$$\frac{\partial^2 k}{\partial l^2}(l,\lambda) = -\frac{f(\lambda)}{4l^{3/2}}$$

and, consequently, $\mathrm{curv}(G) < 0$ on D_p. $\qquad \square$

We will investigate Conjecture 2 with several large deviations arguments. Cramér's theorem about large deviations (Proposition 6 and Exercise 4.3.5, or see [DZ93], for example) applied to Bernoulli random variables yields the following result: Let r be a real number such that $1 \geq r > p$; then

$$\lim_{l \to +\infty} \frac{1}{l} \log \mathbb{P}[S_l \geq rl] = -r \log \frac{r}{p} - (1-r) \log \frac{1-r}{1-p}.$$

Theorem 6 *The large deviations estimate of $\log \mathcal{B}(l,k)$ (Proposition 6) given by*

$$H(l,k) = -k \log \frac{k}{pl} - (l-k) \log \frac{l-k}{(1-p)l}$$

has negative curvature on the domain $\{pl \leq k \leq l\}$.

Proof — The level lines of $H(l,k)$ are defined by

$$k(l,\lambda) \log \frac{k(l,\lambda)}{pl} + (l - k(l,\lambda)) \log \frac{l - k(l,\lambda)}{(1-p)l} = \lambda.$$

We fix λ and we just write $k(l, \lambda) = k(l)$. If we compute the first derivative of the above equation and then simplify, we get

$$k'(l)\log k(l) - k'(l)\log(pl) + (1 - k'(l))\log(l - k(l)) - (1 - k'(l))\log((1 - p)l) = 0.$$

Again, by differentiation,

$$k''(l)\log \frac{(1 - p)k(l)}{p(l - k(l))} - \frac{1}{l} + \frac{k'(l)^2}{k(l)} + \frac{(1 - k'(l))^2}{l - k(l)} = 0,$$

which is equivalent to

$$k''(l)\log \frac{(1 - p)k(l)}{p(l - k(l))} = -\frac{(k(l) - k'(l)l)^2}{lk(l)(l - k(l))}.$$

This last relation shows that $H(l, k)$ has negative curvature on the domain $pl \le k \le l$. $\square$

6.3.3 Proof of Conjecture 1 Under Conjecture 2

Lemma 5 (Under Conjecture 2) If $k - 1 > p(l - 1)$ and $\mu > 0$, then the map

$$x \mapsto \tilde{B}(l + x, k + \mu x)$$

has no local minimum at $x = 0$.

Proof — Call f this map. It is sufficient to prove that either $f'(0) \ne 0$ or $(f'(0) = 0$ and $f''(0) < 0)$. If $f'(0) = 0$, then

$$\mu = -\frac{\tilde{B}_l}{\tilde{B}_k}(l, k),$$

so that

$$f''(0) = \mu^2 \tilde{B}_{kk} + 2\mu \tilde{B}_{kl} + \tilde{B}_{ll} = \mathrm{curv}(\tilde{B})(l, k) \cdot \frac{(\tilde{B}_k^2 + \tilde{B}_l^2)^{3/2}}{\tilde{B}_k^2} < 0$$

thanks to Conjecture 2. $\square$

We now can prove Conjecture 1 under Conjecture 2.

Proof — Because the inequality we want to prove is symmetric in k' and k'', we can suppose that $k''/l'' \ge k'/l'$. If $k + k' - 1 \le p(l + l' - 1)$, then $\tilde{B}(l + l', k + k') > p$ and we have finished. Thus, in the following we assume $k + k' - 1 > p(l + l' - 1)$. Let us define the map

$$f(x) = \tilde{B}(l + x(l' + l''), k + x(k' + k''),) \qquad \text{for} \quad x \in [0, 1].$$

Remark that for $x_0 = l'/(l'+l'') \in (0,1)$,

$$k + x_0(k'+k'') = k + \frac{l'}{l'+l''}(k'+k'') \geq k + \frac{l'}{l'+l''}\left(k' + \frac{k'l''}{l'}\right) = k + k',$$

which implies that $\tilde{\mathcal{B}}(l+l', k+k') \geq f(x_0)$. Hence, it is sufficient to prove that

$$\min\left(p, f(0), f(1)\right) < f(x_0).$$

The set

$$S = \left\{x \in [0,1], \quad k + x(k'+k'') - 1 - p(l + x(l'+l'') - 1) > 0\right\}$$

is a connected segment that contains x_0 because

$$k + x_0(k'+k'') - 1 \geq k + k' - 1 > p(l + l' - 1) = p(l + x_0(l'+l'') - 1).$$

Moreover, S contains 0 or 1 because the linear function involved in the definition of S is either 0 or vanishes only once. By Lemma 5, f has no local minimum on S. We conclude as stated that

$$f(x_0) > \min_{x \in S} f(x) = \min_{x \in \partial S} f(x) \geq \min(p, f(0), f(1)),$$

since if $x \in \partial S \cap (0,1)$, then $f(x) \geq p$ by Lemma 2.   □

Remark: This proof (and the proof of Lemma 5) only relies on the fact that there exists *some* smooth interpolation of the discrete $\mathcal{B}(l,k)$ that has negative curvature on the domain D_p. There are good reasons to think that the $\tilde{\mathcal{B}}(l,k)$ approximation satisfies this property, but it could be that another approximation also does, even though we did not find any (e.g., the piecewise bilinear interpolation of $\mathcal{B}(l,k)$ is not appropriate).

In Figure 6.6, we give the geometric idea underlying the proof of Conjecture 1 under Conjecture 2.

6.3.4 Partial Results About Conjecture 2

In this subsection we will give an asymptotic proof of Conjecture 2. In all of the following we assume that p and r satisfy $0 < p < r < 1$ and $p < 1/2$. The proof relies on the two following technical propositions: Proposition 19 and Proposition 20, which are actually interesting by themselves as they provide a much more precise versions of Hoeffding's inequality.

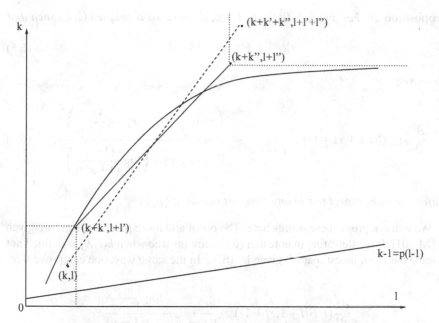

Fig. 6.6 Geometric idea of the proof of Conjecture 1 under Conjecture 2. Assume that $\tilde{\mathcal{B}}(l+l'',k+k'') \leq \tilde{\mathcal{B}}(l+l',k+k')$. Represent the concave level line of $\tilde{\mathcal{B}}$ passing by $(l+l',k+k')$. The point $(l+l'',k+k'')$ is above this level line (indeed, $\frac{\partial \tilde{\mathcal{B}}}{\partial k} < 0$). Since the segments $[(l+l',k+k'),(l+l'',k+k'')]$ and $[(l,k),(l+l'+l'',k+k'+k'')]$ have the same middle point, one of the points (l,k) and $(l+l'+l'',k+k'+k'')$ must lie above the concave level line.

Proposition 19 (Precise large deviations estimate) *Let*

$$D(l+1, rl+1) = \frac{p(1-p)}{(r-p)\sqrt{2\pi lr(1-r)}} \exp\left[-l\left(r\log\frac{r}{p} + (1-r)\log\frac{1-r}{1-p}\right)\right].$$
(6.3)

Then, for any positive p,r,l such that $p < r < 1$ and $p < 1/2$, one has

$$\frac{1 - \dfrac{4r}{(r-p)^2 l(1-p)}}{1 + \dfrac{1}{r(1-r)\sqrt{2\pi lr(1-r)}}} \leq \frac{\tilde{\mathcal{B}}(l+1, rl+1)}{D(l+1, rl+1)} \leq \frac{1}{1 - \dfrac{2}{\sqrt{2\pi lr(1-r)}}}.$$
(6.4)

In particular

$$\tilde{\mathcal{B}}(l+1, rl+1) \underset{l\to+\infty}{\sim} D(l+1, rl+1)$$

uniformly with respect to r in any compact subset of $(p,1)$.

Notice that the exponential term in (6.3) corresponds to Hoeffding's inequality (see Proposition 3).

Proposition 20 *For any $\lambda \in [0,1]$ and $l > 0$, there exists a unique $k(l,\lambda)$ such that*

$$\tilde{\mathcal{B}}(l+1,k(l,\lambda)+1) = \lambda. \tag{6.5}$$

Moreover,

$$\frac{\partial^2 k}{\partial l^2}(l,\tilde{\mathcal{B}}(l+1,rl+1)) \underset{l \to +\infty}{\sim} - \frac{\left(r\log\dfrac{r}{p} + (1-r)\log\dfrac{1-r}{1-p}\right)^2}{l \cdot r(1-r) \cdot \left(\log\dfrac{r(1-p)}{(1-r)p}\right)^3}. \tag{6.6}$$

uniformly with respect to r in any compact subset of $(p,1)$.

We will not prove these results here. The proof and more precise results are given in [Moi01]. It is interesting to note that (6.6) remains true when $k(l,\lambda)$ is defined not from $\tilde{\mathcal{B}}$ but from its estimate D given by (6.3). In the same way, one can prove that

$$\frac{\partial k}{\partial l}(l,\tilde{\mathcal{B}}(l+1,rl+1)) \xrightarrow[l \to +\infty]{} \frac{\log\dfrac{1-p}{1-r}}{\log\dfrac{r(1-p)}{(1-r)p}}$$

is satisfied by both definitions of $k(l,\lambda)$. This proves that (6.3) actually gives a very good estimate of $\tilde{\mathcal{B}}$, since it not only approximates the values of $\tilde{\mathcal{B}}$ but also its level lines up to second order.

Theorem 7 (Asymptotic proof of Conjecture 2) *There exists a continuous map $L : (p,1) \to \mathbb{R}$ such that $(l,k) \mapsto \tilde{\mathcal{B}}(l,k)$ has negative curvature on the domain*

$$D_p^L = \left\{ (l+1,rl+1), \quad r \in (p,1), \quad l \in [L(r),+\infty) \right\}.$$

This result is illustrated on Figure 6.3.4.

Proof — Define $k(l,\lambda)$ by (6.5). Thanks to Proposition 20, the function

$$r \quad \mapsto \quad \frac{\partial^2 k}{\partial l^2}(l,\tilde{\mathcal{B}}(l+1,rl+1)) \cdot \frac{l \cdot r(1-r) \cdot \left(\log\dfrac{r(1-p)}{(1-r)p}\right)^3}{\left(r\log\dfrac{r}{p} + (1-r)\log\dfrac{1-r}{1-p}\right)^2}$$

tends to -1 as l goes to infinity and the convergence is uniform with respect to r in any compact subset of $(p,1)$. Thus, we deduce that the map

$$r \mapsto l(r) = \inf\left\{ l_0 > 0, \, \forall l \ge l_0, \, \mathrm{curv}\,\tilde{\mathcal{B}}(l+1,rl+1) < 0 \right\}$$

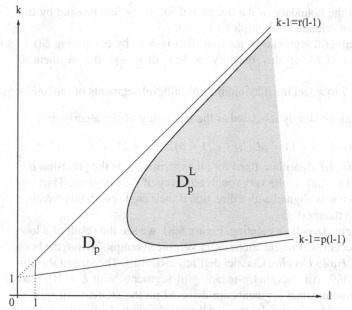

Fig. 6.7 Conjecture 2 is proven on a subdomain D_p^L of D_p.

is bounded on any compact subset of $(p, 1)$. Now, defining $L(r)$ as a continuous upper bound for $l(r)$ yields the desired result. For example, one can take

$$L(r) = \sup_{n \in \mathbb{Z}} d_n(r),$$

where d_n is the unique linear function passing through the points

$$\left(a_{n-1}, \max_{t \in [a_{n-2}, a_n]} l(t) \right) \quad \text{and} \quad \left(a_n, \max_{t \in [a_{n-1}, a_{n+1}]} l(t) \right),$$

and $(a_n)_{n \in \mathbb{Z}}$ is an increasing sequence such that $\lim_{n \to -\infty} a_n = p$ and $\lim_{n \to +\infty} a_n = 1$.
□

6.4 Experimental Results

In all of the experiments, the direction of image pixels is computed on a 2×2 neighborhood with the method described by (5.1) in Chapter 5, with precision $p = 1/16$. The length l of each segment (l, k) is counted in independent pixels, which means that the real length of the segment is $2l$. The number of points having their direction aligned with the direction of the segment with precision p is denoted by k. The algorithm used to find the meaningful segments is the following:

- For each pixel on the boundary of the image and for every line passing by this pixel and having an orientation multiple of $\pi/200$:
- Compute all meaningful segments of the line. This is done by computing $\mathcal{B}(l,k)$ for each segment (l,k). If this quantity is less than $\frac{\varepsilon}{N^4}$, the segment is ε-meaningful.
- Apply Algorithm 7 to select the maximum ε-meaningful segments of the line.

Notice that $\mathcal{B}(l,k)$ can be simply tabulated at the beginning of the algorithm using the relation

$$\mathcal{B}(l+1,k+1) = p\mathcal{B}(l,k) + (1-p)\mathcal{B}(l,k+1).$$

The only parameter of the algorithm, fixed for all experiments, is the precision $p = 1/16$. This value corresponds to the very rough accuracy of 22.5 degrees. Thus two points can be considered as aligned with a direction if their angles with this direction differ by up to ± 22.5 degrees!

In a first experiment (Uccello's painting, Figure 6.8), we see the result of a low-resolution scan of Uccello's "Presentazione della Vergine al tempio" (from the book *L'opera completa di Paolo Uccello*, Classici dell'arte, Rizzoli). The size of the digital image is 467×369. All maximal ε-meaningful segments with $\varepsilon = 10^{-6}$ are displayed. Notice how maximal segments are detected on the staircase in spite of the occlusion by the child going up the steps. All remarks made for Figures 6.1 and 6.2 also apply here. Let us mention in particular the creation by the blur of several parallel alignments at places where a single perceptual alignment is visible.

The experiments of Figures 6.9 and 6.10 first use Algorithm 7 inside each straight line, which keeps only disjoint maximal meaningful segments. Then they apply the general exclusion Algorithm 6 to the whole image and therefore eliminates spurious slanted segments and the effect of blur. The results are self-explanatory.

(a) The original image: Uccello's painting

(b) Maximal ε-meaningful segments for $\varepsilon = 10^{-6}$

Fig. 6.8 Uccello's painting: maximal meaningful alignments. They result from the application of the exclusion principle (Algorithm 7) inside each line. There are many slanted lines detected due to the blur and the low precision. Such slanted alignments can be eliminated by the general exclusion principle.

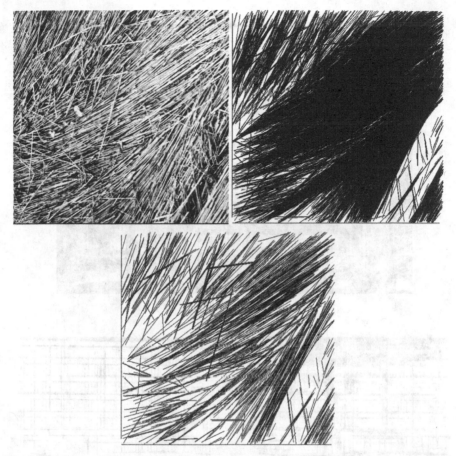

Fig. 6.9 An image of straw: size 512×512. This figure illustrates the accuracy and economy gain in the detection of alignments by applying the exclusion principle. The second image displays all maximal meaningful alignments, obtained by application of Algorithm 7 in each line. Since the image has many parallel alignments, this leads to the detection of many slanted alignments. Indeed, whenever a segment meets many alignments with a small enough angle, it can become meaningful. The way out is given by the general exclusion principle Algorithm 6 which makes each point participate with at most one alignment. Once the exclusion principle has been applied, the set of detected alignments is cleaned up, as displayed in the third image. Notice how two detected alignments may cross.

Notice that the algorithm can also be used for straight contour completion. This is illustrated in Figure 6.8 where the stairs are completed accross the occlusion and in Figure 6.10 where different aligned windows are found on the same maximal meaningful alignment. In the exercise at the end of the chapter, some computations about this property of straight contour completion will be worked out.

(a) A Cachan building

(b) By application of Algorithm 7 on each line, maximal ε-meaningful alignments with $\varepsilon = 10^{-6}$.

(c) Maximal meaningful segments remaining after application of Algorithm 6 extending the exclusion principle.

Fig. 6.10 (a) Building in Cachan. The size of this image is 901×701. (b) All maximal ε-meaningful segments for $\varepsilon = 10^{-6}$. Notice that many slanted maximal alignments are found. The presence of many long and parallel alignments (e.g., at the top of the building) entails the detection of slanted alignments. (c) Segments found after application of the exclusion principle, Algorithm 6.

6.5 Bibliographical Notes

In [GK86], Gerig and Klein have introduced an algorithm for processing an Hough Transform (HT) array before thresholding. The idea of their methods looks very much like the exclusion principle proposed in this chapter. Let us quote Pricen et al. [PIK94] describing the work of Gerig and Klein:

their scheme is quite general and can be applied in cases where a single feature point contributes positive weight to a number of samples in the final array. As a result, it is equally applicable to HT and hypothesis testing. The general idea is that a single edge point should only be a part of one feature in the image. The Gerig and Klein algorithm is a post-processing step that associates each feature point explicitly with the parameter value, out of those to which it contributed a positive weight, which has a maximum value. It can be used for instance to give a precise edge direction to each point by looking along the voting pattern of a point for the minima (maxima) and giving the edge point the appropriate direction. Once this association is made, a new array is reaccumulated using only the single edge direction. The result is an extremely sparse set of peaks in the array. These peaks are processed using a simple 3×3 local maximum detector, and a threshold test is applied.

The exclusion principle used in this chapter was proposed in [DMM04], and a complete study of maximal meaningful segments (first defined in [DMM00]) was developed in [DMM03c]. For a more extended discussion of the links between the exclusion principle, the NFA and variational methods, see Section 15.5 of Chapter 14.

A first implementation of the exclusion principle in the *align-mdl* modulus of the public software MegaWave, due to Andres Almansa and Lionel Moisan, does a simultaneous assignment of every point to the alignment with the best NFA containing it. This simultaneity can lead to the loss of meaningful segments. This drawback can be avoided by an iterative procedure where the most meaningful segment is first computed. Then its points are removed from the other segments, the most meaningful segments among them is computed, and so on. In [Fer06], Fernando Fernandez proposed a fast clever iterative procedure implementing the exclusion principle. He shows that losses of meaningful segments are avoided that way.

6.6 Exercise

6.6.1 Straight Contour Completion

Assume that we have two length l segments on the same line, such that: all of the points belonging to the segments are aligned and that all the points in the length g gap between the segments are not aligned.

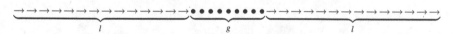

In other words, it means that if A_1 and A_2 denote the two segments, then $\text{NFA}(A_1) = \text{NFA}(A_2) = N^4 \mathcal{B}(l, l, p) = N^4 p^l$. Also, if S_g denotes the whole segment obtained as the union of the segments A_1, A_2, and the gap, then $\text{NFA}(S_g) = N^4 \mathcal{B}(2l + g, 2l, p)$.

The aim of this exercise will be to give conditions on the length g of the gap such that (a) S_g is not a meaningful segment or (b) S_g is more meaningful than A_1 and A_2. In this last case, straight contour completion is achieved.

1) Prove that if $g > 2\frac{1-p}{p}l$ then S_g is not meaningful. Notice that when $p = 1/16$, this condition means $g > 30l$. We will refine this condition in the next question.

2) Let $\alpha(\varepsilon, N^4)$ be the real number defined by $\frac{1}{\sqrt{2\pi}} \int_{\alpha(\varepsilon, N^4)}^{+\infty} e^{-x^2/2} dx = \frac{\varepsilon}{N^4}$. Use Slud's inequality (Theorem 1) to prove that if g is such that

$$p(2l+g) + \alpha(\varepsilon, N^4)\sqrt{(2l+g)p(1-p)} > 2l,$$

then S_g is not ε-meaningful.

3) Let us denote $\mu = g/l$, and $r = 2/(2+\mu)$. Use Hoeffding's inequality (Proposition 3) to prove that if

$$(2+\mu)\left[r\log\frac{r}{p} + (1-r)\log\frac{1-r}{1-p}\right] \geq \log\frac{1}{p},$$

then S_g is maximal meaningful (it is more meaningful than A_1 or A_2).
Numerical values: Prove that if $p = 1/16$, the above condition holds for $\mu = 2$. Use a computer program (e.g., Matlab or Scilab) to prove that the above inequality implies that when $p = 1/16$, then $\mu < \sim 2.2$, and that when $p = 1/8$ then $\mu < \sim 1.3$.

Chapter 7
Modes of a Histogram

7.1 Introduction

The global analysis of digital images can involve the histograms of variables like the gray level or the orientation. Usually histograms are not flat. Peaks and lacunary parts are observed. The peaks can correspond to meaningful groups and the lacunary intervals correspond to separations between them. We will call the peaks *modes of the histogram*. Lacunary intervals are called *gaps*. Since their analysis will in essence be symmetric, we will focus on the modes. How can we decide whether a mode or a gap is meaningful or not? This problem is very similar to the alignment detection problem. As in the meaningful alignment theory, the Helmholtz principle can be adopted. There is indeed no *a priori* knowledge about the histogram model. Thus meaningfulness can be computed as though all samples were uniformly and independently distributed. Meaningful modes will be defined as counterexamples to this uniformity assumption and maximal meaningful modes will be the best counterexamples to uniformity. The *exclusion principle* will be involved again. It will be proven that maximal meaningful modes of the histogram are disjoint. This will give an algorithm that can be immediately applied to image analysis. Can such a detection theory give an account of the so-called "visual pyramid"? According to the visual pyramid doctrine geometric events (gestalts) are grouped recursively at different scales (see Chapter 1). This pyramidal assumption can be confirmed only if the detection of geometric events is robust enough. A first test of *visual pyramid* (i.e., a combination bottom up of gestalt grouping), is given in the last section. All maximal meaningful alignments of an image will be computed and the obtained segments grouped according to the mode of the orientation histogram to which they belong. This yields an implementation of the *parallelism* gestalt.

7.2 Meaningful Intervals

The gray levels of a digital image give a good example of a histogram. The image has M pixels and their values range from 1 to L. In the histogram, the pointwise association of pixels with values is forgotten. For each value a in $\{1,\ldots,L\}$, the histogram only retains the number of points $k(a)$ assuming the gray level a. For each discrete interval of values $[a,b]$, let $k(a,b)$ be the number of points (among the M) with value in $[a,b]$, and let $p(a,b) = (b-a+1)/L$. This represents the prior probability for a point to have value in $[a,b]$.

Histogram modes are defined as intervals $[a,b]$ that contain significantly more points than expected. Adopting the same definition as for alignments, the NFA of an interval $[a,b]$ can be defined as

$$\text{NFA}_1([a,b]) = \frac{L(L+1)}{2} \mathcal{B}(M, k(a,b), p(a,b)), \tag{7.1}$$

where $\mathcal{B}(n,k,p) = \sum_{j=k}^{n} \binom{n}{j} p^j (1-p)^{n-j}$ denotes the tail of the binomial distribution of parameters n and p and where $L(L+1)/2$ is the total number of possible intervals. The NFA is interpreted as in Chapter 5, as the expected number of intervals as meaningful as the one being observed.

Definition 10. *An interval $[a,b]$ is said ε-meaningful if* $\text{NFA}([a,b]) \leq \varepsilon$; *that is,*

$$\mathcal{B}(M, k(a,b), p(a,b)) < \frac{2\varepsilon}{L(L+1)}.$$

Another interpretation already given in Proposition 9 is the following.

Proposition 21 $\text{NFA}_1([a,b])$ *is the smallest value of ε such that $[a,b]$ is ε-meaningful.*

Notice that in Definition 10 (compared to the definition of meaningful alignment), the binomial distribution is used in different ways:

- For histograms: $\mathcal{B}\left(M, k(a,b), \frac{b-a+1}{L}\right)$;
- For alignments: $\mathcal{B}(l,k,p)$.

In the first case, M is fixed. The other arguments, including the probability

$$p(a,b) = \frac{b-a+1}{L} \tag{7.2}$$

depend on $[a,b]$. In the second case, the precision p is fixed and the length l of the segment is a variable first argument of $\mathcal{B}$. All the same, meaningfulness and maximal meaningfulness will receive quite analogous treatment.

Proposition 22 *Let $[a,b]$ be a 1-meaningful interval; then*

$$r(a,b) = \frac{k(a,b)}{M} > p(a,b),$$

and by Hoeffding's inequality,

$$\mathcal{B}(M, k(a,b), p(a,b))$$
$$\leq \exp\left(-M\left[r(a,b)\log\frac{r(a,b)}{p(a,b)} + (1-r(a,b))\log\frac{1-r(a,b)}{1-p(a,b)}\right]\right).$$

Proof .— This is a direct application of Lemma 2 and of Hoeffding's inequality (Proposition 3). Lemma 2 provides the needed inequalities for the binomial distribution only when $p \leq 1/2$. To have the corresponding inequalities for $p > 1/2$ use

$$\mathcal{B}(l,k,p) = 1 - \mathcal{B}(l,l-k+1,1-p)$$

and the relation

$$\mathcal{B}(M,k(a,b),p(a,b)) < \frac{2}{L(L+1)} < \min\left(p(a,b),\frac{1}{2}\right).$$

$\square$

We are interested in a close formula for the detection threshold $k(l)$, namely the minimal number of points making $[a,b]$ 1-meaningful. Let us evaluate how well Hoeffding's inequality performs to give this estimate. In a realistic numerical setting, consider a discrete 256×256 image. The numerical values for M and L are $M = 256^2$ and $L = 256$. The detection threshold $k(l)$ is defined as the smallest integer such that

$$\mathcal{B}\left(M,k(l),\frac{l}{L}\right) < \frac{2}{L(L+1)}.$$

One can also compute the detection thresholds $k_d(l)$ given by the large deviations estimate of the binomial tail (see Proposition 6). This means that $k_d(l)$ is defined as the smallest integer above $M \times l/L$ such that

$$\frac{k_d(l)}{M}\log\frac{k_d(l)L}{Ml} + \left(1 - \frac{k_d(l)}{M}\right)\log\frac{1-k_d(l)/M}{1-l/L} > \frac{1}{M}\log\frac{L(L+1)}{2}.$$

Thanks to Hoeffding's inequality (Proposition 3), $k_d(l) \geq k(l) > Ml/L$.

In Figure 7.1 are plotted $k(l)$, $k_d(l)$ (dotted curve), and $M \times l/L$ (dashed line) for l in $[1,10]$. The maximal value of the relative error $l \mapsto (k_d(l) - k(l))/k(l)$ for $l \in [1,256]$ is about 3% and is attained for small values of l.

These numerical experiments justify adopting the large deviations estimate given by Proposition 6 to define meaningful and maximal meaningful intervals. In practice, $\mathcal{B}(M,k,p)$ is not easily computable when M exceeds 512^2.

Definition 11 (Relative entropy). *The relative entropy of an interval $[a,b]$ (with respect to the prior uniform distribution p) is defined by*

$$H([a,b]) = \begin{cases} 0 & \text{if } r(a,b) \leq p(a,b) \\ r(a,b)\log\frac{r(a,b)}{p(a,b)} + (1-r(a,b))\log\frac{1-r(a,b)}{1-p(a,b)} & \text{otherwise.} \end{cases}$$

If $I = [a,b]$ is an interval with probability $p = p(I)$ and density $r = r(I)$, we note $H(I) = \overline{H}(r(I),p(I)) = \overline{H}(r,p)$.

In the case $r(a,b) > p(a,b)$, the relative entropy $H([a,b])$ is also called the Kullback-Leibler distance between the two Bernoulli distributions of respective parameter $r(a,b)$ and $p(a,b)$ (see [CT91]).

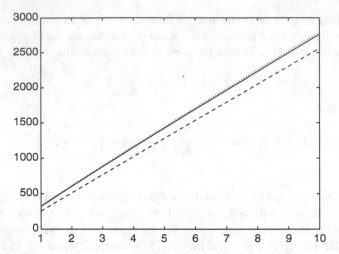

Fig. 7.1 The detection thresholds $k(l)$ and $k_d(l)$ (dotted curve) and Ml/L (dashed line) for $1 \le l \le 10$.

Remark: The above definition can be interpreted in information theory [CT91] as follows. Consider a M points histogram distributed on a length L reference interval. Let I be an interval of length $l \le L$ and $k(I)$ be the number of points it contains. Assume one wants to encode for each point the information of whether it belongs to the fixed interval I or not. Since the prior probability for a point to be in I is l/L, the prior expected bit length needed to encode this information is

$$-k\log_2 \frac{l}{L} - (M-k)\log_2 \left(1 - \frac{l}{L}\right).$$

Indeed, by Shannon's first coding theorem, the optimal average bit length per symbol for encoding the output of a Bernouilli source with parameter p is $-p\log_2(p) - (1-p)\log_2(1-p)$. On the other hand the posterior probability for a point to be in I is k/M. Thus, the posterior expected bit length needed to encode the binarized histogram is

$$-k\log_2 \frac{k}{M} - (M-k)\log_2 \left(1 - \frac{k}{M}\right).$$

The code length gain is the difference between both bit lengths:

$$M\left(r\log_2 \frac{r}{p} + (1-r)\log_2 \frac{1-r}{1-p}\right).$$

where $r = k/M$ and $p = l/L$. Thus, the relative entropy measuring the meaningfulness of an interval is equal to the bit length gain between prior and posterior coding of the interval. The higher the gain, the more meaningful the interval.

Definition 12 (Meaningful interval, large deviation framework). *An interval* $[a,b]$ *is said to be ε-meaningful in the large deviation sense if its relative entropy* $H([a,b])$ *is such that*

$$H([a,b]) > \frac{1}{M} \log \frac{L(L+1)}{2\varepsilon}.$$

From now on we adopt this definition of meaningfulness for intervals. It is slightly less accurate than Definition 10 but quite rewarding in structural properties and computationally affordable.

7.3 Maximal Meaningful Intervals

Maximal meaningful intervals must be introduced for the very same reasons given in Chapter 6 for maximal meaningful alignments.

Definition 13 (Maximal meaningful interval). *An interval* $I = [a,b]$ *is maximal meaningful if it is meaningful and if*

$$\forall J \subset I, \qquad H(J) \leq H(I)$$

$$and \qquad \forall J \supsetneq I, \qquad H(J) < H(I).$$

The natural question then becomes: Does again maximality imply exclusion? In other words, are two maximal meaningful intervals constrained to have no intersection? The following theorem and its corollary assert that, as in the case of alignments, the answer is yes. However, notice that the framework is different. Indeed, in the case of alignments, the probability p was a fixed number and the variables were the length l of the segment and the number k of aligned points on the considered segment. In the case of histograms, the total number of points is a fixed number N and the variables are the prior probability $p(I)$ of interval I and the number $k(I)$ of points in I.

Theorem 8 *Let* I_1 *and* I_2 *be two meaningful intervals such that* $I_1 \cap I_2 \neq \emptyset$; *then*

$$\max(H(I_1 \cap I_2), H(I_1 \cup I_2)) \geq \min(H(I_1), H(I_2))$$

and the inequality is strict when $I_1 \cap I_2 \neq I_1$ *and* $I_1 \cap I_2 \neq I_2$.

Corollary 6. *Let* I *and* J *be two meaningful intervals such that*

$$H(I) = H(J) = \max_{K \subset [0,L]} H(K).$$

Then either $I \subset J$, *or* $J \subset I$, *or* $I \cap J = \emptyset$.

Proof — By Theorem 8, if $I \cap J \neq \emptyset$, $I \not\subset J$ and $J \not\subset I$, then $H(I \cap J)$ or $H(I \cup J)$ exceeds $H(I) = H(J)$, which is a contradiction. $\square$

Proof of Theorem 8 — For an interval I, let $r(I)$ be the proportion of points it contains and $p(I)$ be its relative length. Then the relative entropy of the interval is, according to Definition 11,

$$H(I) = \begin{cases} 0 & \text{if } r(I) \leq p(I) \\ F(r(I), p(I)) & \text{otherwise,} \end{cases}$$

where F is defined on $[0,1] \times [0,1]$ by

$$F(r,p) = r \log r + (1-r) \log(1-r) - r \log p - (1-r) \log(1-p).$$

For all $(r,p) \in [0,1] \times [0,1]$, $F(r,p)$ is positive and it is zero if and only if $r = p$. The partial derivatives of F are

$$\frac{\partial F}{\partial r} = \log \frac{r}{1-r} - \log \frac{p}{1-p} \quad \text{and} \quad \frac{\partial F}{\partial p} = \frac{p-r}{p(1-p)}.$$

$$\frac{\partial^2 F}{\partial r^2} = \frac{1}{r(1-r)}, \quad \frac{\partial^2 F}{\partial r \partial p} = \frac{-1}{p(1-p)} \quad \text{and} \quad \frac{\partial^2 F}{\partial p^2} = \frac{r}{p^2} + \frac{(1-r)}{(1-p)^2}.$$

Thus

$$\frac{\partial^2 F}{\partial r^2} > 0 \quad \text{and} \quad \det(D^2 F) = \frac{\partial^2 F}{\partial r^2} \times \frac{\partial^2 F}{\partial p^2} - \left(\frac{\partial^2 F}{\partial r \partial p} \right)^2 = \frac{(r-p)^2}{r(1-r)p^2(1-p)^2} \geq 0$$

which shows that F is convex. Then the continuous function $\overline{H}(r,p)$ defined by $F(r,p)$ if $r \geq p$ and zero otherwise is also convex (the partial derivatives are continuous). By hypothesis, $I_1 \cap I_2 \neq \emptyset$. Let us denote $I = I_1 \cap I_2$ and $J = I_1 \cup I_2$. Then

$$\begin{cases} r(I) + r(J) = r(I_1) + r(I_2), \\ p(I) + p(J) = p(I_1) + p(I_2) \end{cases} \tag{7.3}$$

and

$$\begin{cases} r(I) \leq \min(r(I_1), r(I_2)) \leq \max(r(I_1), r(I_2)) \leq r(J), \\ p(I) \leq \min(p(I_1), p(I_2)) \leq \max(p(I_1), p(I_2)) \leq p(J). \end{cases} \tag{7.4}$$

We want to show that

$$\min(H(I_1), H(I_2)) \leq \max(H(I), H(J))$$

and that the inequality is strict when $I_1 \cap I_2 \neq I_1$ and $I_1 \cap I_2 \neq I_2$. In the plane $\mathbb{R}^2$, consider the set R of points (r,p) such that $r(I) \leq r \leq r(J)$ and $p(I) \leq p \leq p(J)$. Then R is a rectangle and, by (7.4), it contains the points $X_1 = (r(I_1), p(I_1))$ and $X_2 = (r(I_2), p(I_2))$. Let A be the following set of points:

$$A = \{(r,p) / \overline{H}(r,p) \leq \max(H(I), H(J))\}.$$

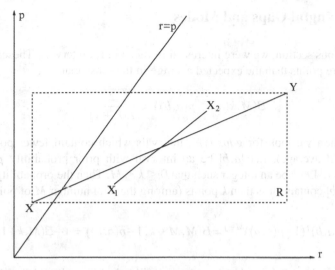

Fig. 7.2 Representation in the plane (r,p) of the intervals I_1, I_2, $I = I_1 \cap I_2$, and $J = I_1 \cup I_2$.

We just have to prove that A contains one of I_1 and I_2. A is a convex set because H is a convex function. Let $X = (r(I), p(I))$ and $Y = (r(J), p(J))$; then A contains the segment $[X, Y]$. Since $\frac{\partial F}{\partial r} = \frac{\partial H}{\partial r} \geq 0$ for $r \geq p$, the set A contains $R \cap \{r \geq p\} \cap \mathcal{P}_+$, where $\mathcal{P}_+$ is the half-plane above the line (X, Y) (see Figure 7.2.)

I_1 and I_2 being meaningful, the points X_1 and X_2 belong to $R \cap \{r > p\}$. Since the middle point of segment $[X_1, X_2]$ also is the middle point of segment $[X, Y]$ by (7.3), one of X_1 and X_2 belongs to $\mathcal{P}_+$. Thus, X_1 or X_2 belongs to A, which shows that

$$\min(H(I_1), H(I_2)) \leq \max(H(I), H(J)).$$

If $I \neq I_1$ and $I \neq I_2$, then the inequality is strict, due to the fact that for $r > p$, $\frac{\partial F}{\partial r} > 0$. □

Proposition 23 *Let I_1 and I_2 be two different maximal meaningful intervals; then*

$$I_1 \cap I_2 = \emptyset.$$

Proof — Assume that $I = I_1 \cap I_2 \neq \emptyset$. If $I \neq I_1$ and $I \neq I_2$ then by Theorem 8

$$\max(H(I_1 \cap I_2), H(I_1 \cup I_2)) > \min(H(I_1), H(I_2)),$$

which is a contradiction with the fact that I_1 and I_2 are maximal meaningful. If, for example, $I = I_1 \cap I_2 = I_1$, then $I_1 \subset I_2$. Since, by hypothesis, I_1 and I_2 are maximal meaningful, we get, by definition of the maximality, $H(I_1) \leq H(I_2)$ and $H(I_2) < H(I_1)$, which is again a contradiction. □

7.4 Meaningful Gaps and Modes

In the previous section, we were interested in meaningful intervals. These intervals contain more points than the expected average in the sense that

$$\mathcal{B}(M,k(a,b),p(a,b)) < \frac{2}{L(L+1)}.$$

In this section we look for *gaps* (i.e., intervals which contain fewer points than the expected average). Let $[a,b]$ be an interval with prior probability $p(a,b) = (b-a+1)/L$. Let k be an integer such that $0 \leq k \leq M$. Then the probability that the interval $[a,b]$ contains less than k points (among the total number M of points) is

$$\sum_{j=0}^{k} \binom{M}{j} p(a,b)^j (1-p(a,b))^{M-j} = \mathcal{B}(M,M-k,1-p(a,b)) = 1-\mathcal{B}(M,k+1,p(a,b)).$$

An interval $[a,b]$ containing $k(a,b)$ points is said to be a meaningful gap if

$$\mathcal{B}(M,M-k(a,b),1-p(a,b)) < \frac{2}{L(L+1)}.$$

Proposition 24 *An interval cannot be at the same time a meaningful interval and a meaningful gap.*

Proof — Let $[a,b]$ be a meaningful gap; then, as in the proof of Proposition 22, thanks to Lemma 2,

$$M-k(a,b) > M \cdot (1-p(a,b)),$$

(i.e., $r(a,b) = k(a,b)/M < p(a,b)$). This shows that $[a,b]$ cannot be a meaningful interval.   □

This result shows the consistency of the definition of gaps and modes. From now on, and by exactly the same arguments as in the previous section, the large deviation estimate will be adopted for defining meaningful gaps.

Definition 14 (Meaningful gap). *An interval $[a,b]$ containing $k(a,b)$ points is said to be a meaningful gap (in the large deviations framework) if and only if $r(a,b) = k(a,b)/M < p(a,b)$ and*

$$r(a,b) \log \frac{r(a,b)}{p(a,b)} + (1-r(a,b)) \log \frac{1-r(a,b)}{1-p(a,b)} > \frac{1}{M} \log \frac{L(L+1)}{2}.$$

In the large deviation framework, it follows immediately from Definitions 14 and 12 that an interval cannot be both a gap and a mode.

Definition 15 (Meaningful mode). *An interval is a meaningful mode if it is a meaningful interval and if it does not contain any meaningful gap.*

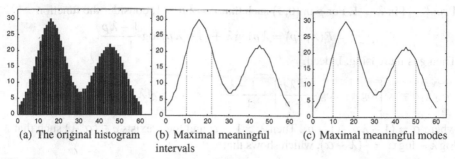

    (a) The original histogram   (b) Maximal meaningful   (c) Maximal meaningful modes
                                  intervals

Fig. 7.3 Comparison between maximal meaningful intervals and maximal meaningful modes.

Definition 16 (Maximal meaningful mode). *An interval I is a maximal meaningful mode if it is a meaningful mode and if for all meaningful modes $J \subset I$, $H(J) \leq H(I)$ and for all meaningful modes $J \supsetneq I$, $H(J) < H(I)$.*

When $I \subset J$ and if I and J are equally meaningful $(H(I) = H(J))$, the larger interval J is maximal meaningful. Figure 7.4 illustrates the difference between meaningful intervals and modes. Figure 7.4(a) is the original histogram with $L = 60$ and $M = 920$. Figure 7.4(b) shows that this histogram has only one maximal meaningful interval: the interval $[10, 22]$. The second peak, $[40, 50]$, is not maximal meaningful because

$$\text{NFA}([10, 22]) < \text{NFA}([10, 50]) < \text{NFA}([40, 50]).$$

In Figure 7.4(c) two maximal meaningful modes are obtained: the intervals $[10, 22]$ and $[40, 50]$.

7.5 Structure Properties of Meaningful Intervals

7.5.1 Mean Value of an Interval

The aim here is to compare the relative entropy of two intervals that have the same mean value. The normalized mean value of an interval $[a, b]$ is defined as $r(a, b)/p(a, b)$, where $r(a, b) = \frac{k(a,b)}{M}$ and $p(a, b) = \frac{b-a+1}{L}$. Notice that the normalized mean value of a meaningful interval is always larger than 1.

Proposition 25 *Let I and J be two intervals with the same mean value*

$$\lambda = \frac{r(I)}{p(I)} = \frac{r(J)}{p(J)} > 1.$$

If $p(I) > p(J)$, then

$$H(I) > H(J).$$

In other words, when the average is fixed, the more meaningful interval is the longer one.

Proof — Fix $\lambda > 1$. For p in $(0,1)$ such that $r = \lambda p \leq 1$, consider the function

$$g(p) = F(\lambda p, p) = \lambda p \log \lambda + (1 - \lambda p) \log \frac{1 - \lambda p}{1 - p}.$$

Then g is increasing. Indeed,

$$g'(p) = \lambda \left[\log \lambda - \log \frac{1 - \lambda p}{1 - p} \right] + \frac{1 - \lambda p}{1 - p} - \lambda = \lambda \left[\log \lambda - \log \alpha \right] - (\lambda - \alpha),$$

where $\alpha = (1 - \lambda p)/(1 - p)$. Then $\lambda > 1 > \alpha$ and there exists $c \in (\alpha, \lambda)$ such that $\log \lambda - \log \alpha = \frac{1}{c}(\lambda - \alpha)$, which shows that

$$g'(p) = \frac{\lambda}{c}(\lambda - \alpha) - (\lambda - \alpha) > 0.$$

$\square$

Corollary 7. *Let $[a,b]$ and $[b+1,c]$ be two consecutive meaningful intervals, then*

$$H(a,c) \geq \min[H(a,b), H(b+1,c)].$$

In other words, $[a,c]$ is more meaningful than at least one of $[a,b]$ and $[b+1,c]$.

Proof — Since $r(a,c) = r(a,b) + r(b+1,c)$ and $p(a,c) = p(a,b) + p(b+1,c)$, we get

$$\frac{r(a,c)}{p(a,c)} \geq \min \left[\frac{r(a,b)}{p(a,b)}, \frac{r(b+1,c)}{p(b+1,c)} \right]$$

and then the result is a direct consequence of the previous proposition. $\square$

As a consequence, maximal meaningful intervals cannot be consecutive.

7.5.2 Structure of Maximal Meaningful Intervals

Theorem 9 *Let h be a histogram defined on a finite set of values $\{1, ..., L\}$. If $[a,b]$ is a maximal meaningful interval such that $1 < a < b < L$, then*

$$h(a - 1) < h(a) \quad and \quad h(b + 1) < h(b),$$
$$h(a) > h(b + 1) \quad and \quad h(b) > h(a - 1).$$

Figure 7.4 illustrates the theorem by presenting the typical structure of a maximal meaningful interval. These compatibility conditions for the endpoints of maximal meaningful interval permit one to speed up the computation.

Proof — Let $M = \sum_{i=1}^{L} h(i)$ be the total mass of the histogram. For each interval $[i, j]$

$$p(i, j) = \frac{j - i + 1}{L} \quad and \quad r(i, j) = \frac{\sum_{x=i}^{j} h(x)}{M}.$$

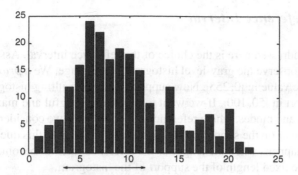

Fig. 7.4 Maximal meaningful interval of a discrete histogram.

The relative entropy $H([i,j]) = \overline{H}(r(i,j), p(i,j))$ of the interval $[i,j]$ is 0 if

$$r(i,j) < p(i,j)$$

and

$$r(i,j)\log r(i,j) + (1-r(i,j))\log(1-r(i,j)) - r(i,j)\log p(i,j)$$
$$+ (1-r(i,j))\log(1-p(i,j))$$

otherwise. The proof of the theorem mainly uses the fact that the function $(r,p) \mapsto \overline{H}(r,p)$ is convex (see the proof of Theorem 8) and that $\frac{\partial \overline{H}}{\partial r} \geq 0$ for $r \geq p$.

Let $[a,b]$ be a maximal meaningful interval. We shall prove that $h(a-1) < h(a)$ (the proof is exactly the same for the other inequalities). Assume by contradiction that $h(a-1) \geq h(a)$. Since $[a,b]$ is a meaningful interval, then $r(a,b) > p(a,b)$. Using the strict convexity of $\overline{H}(r,p)$ for $r > p$,

$$H(a,b) < \max\left(\overline{H}\left(r(a,b) - \frac{h(a)}{M}, p(a,b) - \frac{1}{L}\right), \overline{H}\left(r(a,b) + \frac{h(a)}{M}, p(a,b) + \frac{1}{L}\right)\right).$$

Since $[a,b]$ is maximal,

$$H(a+1,b) = \overline{H}\left(r(a,b) - \frac{h(a)}{M}, p(a,b) - \frac{1}{L}\right) \leq H(a,b).$$

Thus,

$$H(a,b) < \overline{H}\left(r(a,b) + \frac{h(a)}{M}, p(a,b) + \frac{1}{L}\right).$$

This shows that $r(a,b) + h(a)/M > p(a,b) + 1/L$. Using the fact that $\frac{\partial \overline{H}}{\partial r} \geq 0$ for $r \geq p$, one gets

$$H(a-1,b) = \overline{H}\left(r(a,b) + \frac{h(a-1)}{M}, p(a,b) + \frac{1}{L}\right) \geq \overline{H}\left(r(a,b) + \frac{h(a)}{M}, p(a,b) + \frac{1}{L}\right).$$

Thus,

$$H(a-1,b) > H(a,b),$$

which is a contradiction with the maximality of $[a,b]$. $\square$

7.5.3 The Reference Interval

The problem addressed here is the choice of the reference interval. Assume, for example, that we observe the gray-level histogram of an image. We *a priori* know that gray levels have value in $[0, 255]$. Now suppose that the resulting histogram has, for example, support in $[50, 100]$. If we want to detect meaningful and maximal meaningful intervals and modes, which reference interval should we consider? Should we work on $[0, 255]$ or on the support of the histogram? To answer this question we can first ask what happens when the length of the reference interval becomes very large compared to the fixed length of the support of the histogram.

Let h be a discrete histogram *a priori* defined on a finite set of values $\{1, ..., L\}$. Assume that the support of the histogram is $[1, n]$ (i.e., $h(1) > 0$, $h(n) > 0$ and $h(x) = 0$ for $x > n$). For a discrete interval $[a, b] \subset [1, n]$, let $H_L([a, b])$ denote its relative entropy when the reference interval is $[1, L]$ and let $H([a, b])$ denote its relative entropy when the reference interval is $[1, n]$ (i.e., the support of the histogram).

Proposition 26 *Let h be a discrete histogram with support $[1, n]$. Let L be the length of the reference interval. Then there exists L_0 such that*

$$\forall L \geq L_0, \quad \forall [a, b] \neq [1, n], \quad H_L([a, b]) < H_L([1, n]).$$

This means that when the length of the reference interval is large enough, the support of a discrete histogram is maximal meaningful (and it is the only one).

Proof — For a discrete interval $[a, b] \subset [1, n]$, let $p(a, b)$ denote its relative length and $r(a, b)$ its relative weight when the reference interval is the support $[1, n]$. Let $p_L(a, b)$ denote its relative length and $r_L(a, b)$ its relative weight when the reference interval is the support $[1, L]$. We then have

$$p_L(a, b) = \frac{n}{L} p(a, b) \quad \text{and} \quad r_L(a, b) = r(a, b).$$

Thus,

$$H_L([a, b]) = H([a, b]) + r(a, b) \log \frac{L}{n} + (1 - r(a, b)) \log \frac{1 - p(a, b)}{1 - np(a, b)/L}. \tag{7.5}$$

In particular, $H_L([1, n]) = \log(L/n)$ and the last term of (7.5) being negative (because $L \geq n$), we get

$$H_L([a, b]) \leq H([a, b]) + r(a, b) \log \frac{L}{n}.$$

If $[a, b] \neq [1, n]$, then $1 - r(a, b) > 0$. Consequently, there exists a constant C such that

$$\forall [a, b] \neq [1, n], \quad \frac{H([a, b])}{1 - r(a, b)} < C.$$

This shows that for all L such that $\log(L/n) > C$, then $H_L([a, b]) < H_L([1, n])$ for all $[a, b] \neq [1, n]$. $\qquad\square$

(a) The original road image

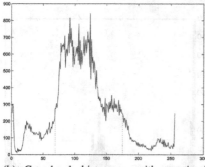

(b) Grey-level histogram with maximal meaningful interval

(c) Quantized Image

Fig. 7.5 Maximal meaningful modes and optimal gray level quantization of a digital image. (a) Original image; (b) histogram of gray-levels of the image with its single maximal meaningful interval $[69, 175]$ (between the dotted lines); (c) quantized image: black points represent points in the original image with gray level in $[0, 68]$, gray points represent points with gray level in the maximal meaningful interval $[69, 175]$ and white points represent points with gray level larger than 176.

7.6 Applications and Experimental Results

A first application is the study of an image gray-level histogram. Looking for the maximal meaningful intervals is a way to obtain a gray-level quantization (Figure 7.5.) This section also illustrates some joint applications of meaningful alignments and modes. The ultimate goal of Computer Vision systems is to combine elementary detection modules as the ones we defined in this book into more and more sophisticated geometric interpretation. Being able to combine alignment detection with parallelism detection is a sanity check. The maximal meaningful alignments of a digital image are a finite set of segments. Each one has a length and an orientation (valued in $[0, 2\pi)$ because segments are oriented). The precision of the direction of the segment is related to its length: If l denotes the length of the segment, measured as usual with the pixel size as unit, the precision of its direction is $1/l$.

(a) The original image: Uccello's painting

(b) Maximal ε-meaningful segments for $\varepsilon = 10^{-6}$

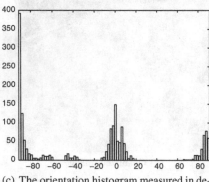

(c) The orientation histogram measured in degrees

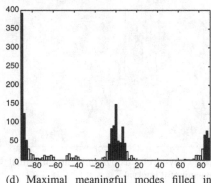

(d) Maximal meaningful modes filled in black

Fig. 7.6 Uccello's painting: maximal meaningful alignments and histogram of orientations. Two maximal meaningful modes are found corresponding respectively to the horizontal and vertical segments.

To build the discrete histogram of the orientations of detected alignments, the interval $[0, 2\pi)$ is decomposed into $n = 2\pi l_{min}$ bins, where l_{min} is the minimal length of the detected segments. Thus, the size of a bin is $1/l_{min}$. One can then compute the maximal meaningful modes of this orientation histogram. The framework is slightly different from what we have defined in this chapter. Indeed, a histogram of orientations is defined on the "circular" interval $[0, 2\pi)$. Thus intervals $[a, b]$ can have $0 \leq a \leq b < 2\pi$, but also $0 \leq b \leq a < 2\pi$. An interval $[a, b]$ such that $0 \leq b \leq a < 2\pi$ is defined as the union $[a, 2\pi) \cup [0, b]$. This does not alter the validity of the meaningfulness definitions, provided the number of considered intervals in the definition of the NFA is changed accordingly.

A first example is treated in Figure 7.6. In Figure 7.6(b), all the 2925 maximal ε-meaningful segments with $\varepsilon = 10^{-6}$ are displayed. In Figure 7.6(c) the histogram of the orientations modulo π is shown. The orientation is measured in degrees and the $[-90, 90]$ degrees interval is divided into 85 bins. In Figure 7.6(d) one can see the maximal meaningful modes of the orientation histogram (filled in black). There are two maximal meaningful modes: $[-6.5; 8.5]$ (corresponding to all the horizontal segments) and $[85; -85]$ (corresponding to the vertical ones).

7.7 Bibliographic Notes

In histogram analysis there are several classes of algorithms computing modes. First, a parametric model can be at hand, ensuring, for example, that the histogram is an instance of k Gaussian random variables whose average and variance have to be estimated from the histogram ([DH73], [TC92], [PHB99]). Optimization algorithms can be defined for this problem and, if k is unknown, it may be found by using variants of the Minimal Description Length Principle [Ris89]. Many theories intend to threshold a histogram in an optimal way (i.e., to divide the histogram into two modes according to some criterion). The most popular criterion is based on entropy (see [Pun81], [Abu89], [KSW85], [CCJA94]). These authors try to find a threshold value m such that the entropy of the bimodal histogram is maximal. An obvious generalization finds multiple thresholds by entropy criteria. This threshold problem turns out to be very useful and relevant in image analysis, since it leads to the problem of optimal quantization of the gray levels. However, here again, the proposed thresholds are not proved to be relevant nor to separate meaningful modes of the histogram. To take an instance, if the histogram is constant, the optimal threshold given by the mentioned methods is the median value. Now, a constant histogram is not bimodal.

Generalizations (in particular, recursive applications) of the detection of meaningful modes and gaps of an histogram have been developed by Delon, Desolneux, Lisani, and Petro in [DDLP07b] (also in [DDLP04] with application to fast camera stabilization) and in [DDLP07a] (with application to an automatic color palette).

See also the exercises of Section 7.8 for more references in statistics about histograms (in particular about Grenander estimator and unimodal densities estimation).

7.8 Exercises

7.8.1 Kullback-Leibler Distance

Let P and Q be two discrete probability distributions on the finite set $\Omega_n = \{1, 2, \ldots, n\}$. The *relative entropy* or *Kullback-Leibler distance* between P and Q is defined by

$$\mathrm{KL}(P\|Q) = \sum_{k=1}^{n} P(k) \log \frac{P(k)}{Q(k)}.$$

1) Use an inequality of convexity to prove that $\mathrm{KL}(P||Q) \geq 0$ and that $\mathrm{KL}(P||Q) = 0$ if and only if $P = Q$. Notice that, in general, one has $\mathrm{KL}(P||Q) \neq \mathrm{KL}(Q||P)$. Hence, the Kullback-Leibler distance is not a "true" distance.

2) Prove that the function $(P,Q) \mapsto \mathrm{KL}(P||Q)$ is a convex function.

3) Let $a_1, \ldots, a_m$ and $b_1, \ldots, b_m$ be non-negative numbers. Prove the following log-sum inequality:

$$\sum_{i=1}^{m} a_i \log \frac{a_i}{b_i} \geq \left(\sum_{i=1}^{m} a_i \right) \log \frac{\sum_{i=1}^{m} a_i}{\sum_{i=1}^{m} b_i}.$$

4) Let p and q be two real numbers in $[0,1]$ such that $p \geq q$. Prove that

$$p \log \frac{p}{q} + (1-p) \log \frac{1-p}{1-q} \geq 2(p-q)^2.$$

(*Hint:* Consider the difference of the two terms and compute the partial derivative with respect to q).

5) Let the L^1-distance between the two probability distributions P and Q be defined by

$$\| P - Q \|_1 = \sum_{k=1}^{n} |P(k) - Q(k)|.$$

Let $\Omega_+ = \{k \,|\, P(k) > Q(k)\}$ and let $p_+ = \sum_{k \in \Omega_+} P(k)$ and $q_+ = \sum_{k \in \Omega_+} Q(k)$. Prove first that

$$\| P - Q \|_1 = 2(p_+ - q_+).$$

Then prove (using questions 3 and 4) that

$$\mathrm{KL}(P||Q) \geq \frac{1}{2} \| P - Q \|_1^2.$$

Many results about the Kullback-Leibler distance and links to Information Theory can be found in the book of Cover and Thomas [CT91].

7.8.2 A Qualitative a Contrario Hypothesis

Let r be a discrete observed normalized histogram on $\{1, \ldots, L\}$. Instead of asking "does this histogram have meaningful modes or gaps according to the a-contrario uniform hypothesis?" which was the question addressed in this chapter, we want to ask: "is this histogram meaningfully decreasing?", in other words: "does it have meaningful gaps or modes according to the a-contrario hypothesis of decrease?"

Let us first introduce some notations. Let $\mathcal{P}(L)$ denote the space of probability distributions on $\{1, 2, \ldots, L\}$ (i.e., the vectors $p = (p_1, \ldots, p_L)$ such that $\forall i, p_i \geq 0$ and $\sum_{i=1}^{L} p_i = 1$). Let $\mathcal{D}(L)$ denote the space of probability distributions on $\{1, 2, \ldots, L\}$ that have the property of being decreasing: $p_i \geq p_{i+1}$ for all $i \leq L - 1$.

1) Prove that there exists a unique $\bar{r} \in \mathcal{D}(L)$ that achieves the minimal Kullback-Leibler distance from r to $\mathcal{D}(L)$; that is,

$$KL(r||\bar{r}) = \min_{p \in \mathcal{D}(L)} KL(r||p),$$

where $\forall p \in \mathcal{D}(L)$, $KL(r||p) = \sum_{i=1}^{L} r_i \log(r_i/p_i)$.

The distribution $\bar{r}$ is known as the Grenander estimator of r. Introduced by Grenander ([Gre80]), this estimator is defined as the non-parametric maximum likelihood estimator restricted to decreasing densities on the line. Grenander shows in [Gre80] (see also [BBBB72]) that $\bar{r}$ is merely "the slope of the smallest concave majorant function of the empirical repartition function of r".

2) Pool Adjacent Violators Algorithm

Let $r = (r_1, ..., r_L) \in \mathcal{P}(L)$ be a normalized histogram. We consider the operator $D : \mathcal{P}(L) \to \mathcal{P}(L)$ defined by the following: For $r \in \mathcal{P}(L)$, and for each interval $[i, j]$ on which r is increasing (i.e., $r_i \leq r_{i+1} \leq \cdots \leq r_j$ and $r_{i-1} > r_i$ and $r_{j+1} < r_j$), we set

$$D(r)_k = \frac{r_i + ... + r_j}{j - i + 1} \quad \text{for } k \in [i, j], \quad \text{and} \quad D(r)_k = r_k \text{ otherwise.}$$

This operator D replaces each increasing part of r by a constant value (equal to the mean value on the interval). Prove that after a finite number (less than the size L of r) of iterations of D we obtain the decreasing distribution $\bar{r}$:

$$\bar{r} = D^L(r).$$

The above algorithm is called "Pool Adjacent Violators Algorithm" (see [ABE+55], [Bir89] for more details). Its generalization to unimodal densities estimation can be found in [Bir97].

3) Prove that $\bar{r}$ also achieves the minimal L^2-distance from r to $\mathcal{D}(L)$.

The definitions of meaningful intervals and gaps for the a-contrario decreasing hypothesis are analogous of the ones given in this chapter: We just replace the uniform distribution by the decreasing distribution $\bar{r}$. This leads to the following definition:

Definition 17. *Let r be an observed normalized histogram. We say that an interval $[a, b]$ is meaningful for the decreasing hypothesis (resp. a meaningful gap for the decreasing hypothesis) if $r(a, b) > \bar{r}(a, b)$ (resp. $r(a, b) < \bar{r}(a, b)$) and*

$$H_{\bar{r}}([a, b]) \geq \frac{1}{M} \log \frac{L(L+1)}{2},$$

where M is the number of samples and

$$H_{\bar{r}}([a, b]) = r(a, b) \log \frac{r(a, b)}{\bar{r}(a, b)} + (1 - r(a, b)) \log \frac{1 - r(a, b)}{1 - \bar{r}(a, b)}.$$

Chapter 8
Vanishing Points

8.1 Introduction

Sets of parallel lines in 3-D space are projected into a 2-D image to a set of concurrent lines. The meeting point of these lines in the image plane is called a *vanishing point*. It can belong to the line at infinity of the image when the 3-D lines are parallel to the image plane. Even though concurrence in the image plane does not necessarily imply parallelism in 3-D, the counterexamples for this implication are rare in real images. Thus, the problem of grouping sets of parallel lines in 3-D is reduced to finding significant sets of concurrent lines in the image plane.

In this chapter we will explain how vanishing points can be reliably and automatically detected with a low number of false alarms (NFA) and a high precision level without using any *a priori* information on the image or calibration parameters and without any parameter tuning. On the contrary, a vanishing point detector can be used to determine some calibration parameters of the camera or for other applications such as single-view metrology.

8.2 Detection of Vanishing Points

As in the case of alignments, a meaningful vanishing point can be defined in terms of the Helmholtz principle. The objects undergoing grouping will be the meaningful segments (Chapter 5) obtained by the final method presented in Chapter 6. The common property of some of these segments is a common point v_∞ met by their supporting lines.[1] Due to measurement errors, there will never be a large number of segments intersecting at a single point v_∞; rather, there will be a family of lines meeting a more or less small subset V of the image plane, which will be called the

---

[1] If the segments are pinhole projections of 3-D lines with a common orientation d_∞, then this point v_∞ is the pinhole projection of the 3-D orientation d_∞.

133

vanishing region. To consider all possibilities, we need to define a *finite* family of such regions $\{V_j\}_{j=1}^M$ covering the whole (*infinite*) image plane, that is,

$$\bigcup_{j=1}^{M} V_j = \mathbb{R}^2. \tag{8.1}$$

8.2.1 Meaningful Vanishing Regions

Let N denote the final number of meaningful segments detected in the image after one has applied the exclusion principle (Chapter 6). Here, the Helmholtz principle is adapted to the case in which the objects that are observed are the supporting lines $l_1, l_2, \ldots, l_N$ of the N line segments obtained. The common property tested for is whether a group of k such lines intersects one of the vanishing regions V_j. Figure 8.1(a) illustrates this construction. Under the assumption that all lines are independent with the same distribution, the probability of such an event is $\mathcal{B}(N, k, p_j)$, where p_j is the probability that a line meets the vanishing region V_j. Moreover, since the M regions V_j are chosen to sample all possible vanishing regions, we are making

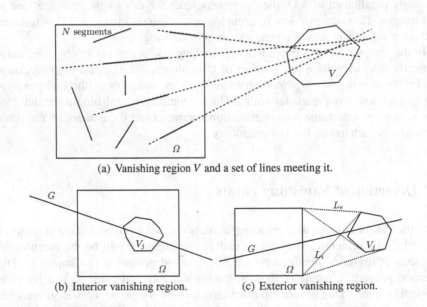

(a) Vanishing region V and a set of lines meeting it.

(b) Interior vanishing region.   (c) Exterior vanishing region.

Fig. 8.1 Meaningful vanishing regions. (a) The problem consists of estimating the expected number of occurrences of the event "at least k out of N lines meet a vanishing region V_j," given that the lines meet (are visible in) the image domain Ω. In order to compute the associated probabilities $p = \mathbb{P}[G \text{ meets } V_j \,|\, G \text{ meets } \Omega]$ we distinguish two cases: (b) interior vanishing regions $V_j \subseteq \Omega$: in this case, $p = \text{Per}(V_j)/\text{Per}(\Omega)$; (c) exterior vanishing regions $V_j \cap \Omega = \emptyset$: in this case, $p = (L_i - L_e)/\text{Per}(\Omega)$.

$N_T = M$ such tests. Thus, the number of false alarms for a vanishing region V_j can be defined as

$$\text{NFA}(V_j) = M \, \mathcal{B}(N, k, p_j), \tag{8.2}$$

and, as usual, the vanishing region is ε-meaningful if k is sufficiently large to have $\text{NFA}(V_j) \le \varepsilon$.

In order to actually find the value of NFA and the value of the minimal value $k(j, \varepsilon)$ of k such that V_j becomes meaningful, the probabilities p_j have to be computed. This is the subject of the next subsection.

8.2.2 Probability of a Line Meeting a Vanishing Region

Up to this point, the analysis has been formally almost equivalent to the case of meaningful alignments, stressing the duality of the events "n lines meet at a point" and "n points belong to a single line." The threshold $k(j, \varepsilon)$ is computed in the same manner and from the same binomial tail as in the alignment case. The only difference is the interpretation of the parameters. Here the total number of segments N plays the role of the length of the segment in the case of alignments. The number of events M represents the total number of possible vanishing regions (instead of the total number of possible segments). The specific geometry of the vanishing point problem comes into play only at this point in computing the probability p_j that a random line of the plane hits simultaneously the image and a vanishing region V_j.

Thankfully, this geometric probability problem can be elegantly solved (see [San76]), yielding a closed-form formula in terms of the internal and external perimeters of both regions. Here we state the main result from integral geometry that is needed to compute p_j. We refer the reader to the exercises at the end of the chapter for the proofs and to the treatise by Santalo [San76] for a complete development of the theory leading to this and other interesting results in stochastic geometry.

First, the polar coordinates parameterization for a random line on the plane G is considered (see Figure 8.2(a)):

$$G(\rho, \theta) = \{(x,y) \in \mathbb{R}^2 : x \cos\theta + y \sin\theta = \rho\}, \tag{8.3}$$

and it is shown (see Exercise 8.5.1) from symmetry arguments that the only translation and rotation invariant measure for sets of lines is $dG = d\rho \, d\theta$ (up to an irrelevant multiplicative constant). The main result from integral geometry that will be used here is the following (see Exercise 8.5.3).

Proposition 27 *Consider two convex sets K_1 and K_2 of the plane. Assume that they are bounded, closed, and with nonempty interior. Let $\text{Per}(K_1)$ and $\text{Per}(K_2)$ denote their respective perimeters. Then the measure of all lines meeting both sets is*

$$\mu[G \cap K_1 \ne \emptyset \text{ and } G \cap K_2 \ne \emptyset] = \begin{cases} \text{Per}(K_1) & \text{if } K_1 \subseteq K_2 \\ L_i - L_e & \text{if } K_1 \cap K_2 = \emptyset \quad (8.4) \\ \text{Per}(K_1) + \text{Per}(K_2) - L_e & \text{otherwise,} \end{cases}$$

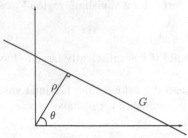

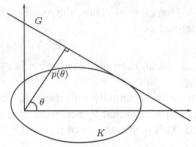

(a) Parameterization of a random line $G(\rho,\theta)$ in the plane.

(b) Support function $p(\theta)$ of a convex set.

Fig. 8.2 (a) With the line parameters (ρ,θ) illustrated in this figure, the measure $dG = d\rho\,d\theta$ becomes translation and rotation invariant. (b) The support function $p(\theta)$ defined in Equation (8.13) uniquely determines the convex set K and plays a central role in computing the probability that a line meets a convex set K.

where the external perimeter L_e is the perimeter of the convex hull of K_1 and K_2 and the internal perimeter L_i is the length of the "internal envelope" of both sets, which is composed of the internal bitangents to K_1 and K_2 and parts of their perimeters. Figure 8.1(c) illustrates this construction for $K_1 = \Omega$ and $K_2 = V_j$.

This result can be directly applied to the problem of determining p_j in the case that the vanishing region $V_j \subseteq \Omega$ is contained in the (convex) image domain Ω. Since the only observed line segments meet the image domain, the probability we are interested in is actually

$$p_j = \mathbb{P}[G \cap V_j \neq \emptyset \mid G \cap \Omega \neq \emptyset]$$
$$= \frac{\mu[G \cap V_j \neq \emptyset \text{ and } G \cap \Omega \neq \emptyset]}{\mu[G \cap \Omega \neq \emptyset]}$$
$$= \frac{\mu[G \cap V_j \neq \emptyset]}{\mu[G \cap \Omega \neq \emptyset]} \tag{8.5}$$
$$= \frac{\text{Per}(V_j)}{\text{Per}(\Omega)}.$$

For vanishing regions external to the image domain ($V_j \cap \Omega = \emptyset$), using the second case of (8.4), the probability becomes

$$p_j = \mathbb{P}[G \cap V_j \neq \emptyset \mid G \cap \Omega \neq \emptyset]$$
$$= \frac{\mu[G \cap V_j \neq \emptyset \text{ and } G \cap \Omega \neq \emptyset]}{\mu[G \cap \Omega \neq \emptyset]} \tag{8.6}$$
$$= \frac{L_i - L_e}{\text{Per}(\Omega)}.$$

Note that the intermediate case, in which there is an intersection but no inclusion, is treated as this second case with $L_i = \text{Per}(K_1) + \text{Per}(K_2)$, which is true in the limiting case when K_2 is tangent but exterior to K_1. In the other limiting case, when K_2 is tangent but interior to K_1, we still have $L_i = \text{Per}(K_1) + \text{Per}(K_2)$, but $L_e = \text{Per}(K_1)$, so $L_i - L_e = \text{Per}(K_2)$ and we are back to the first case.

8.2.3 Partition of the Image Plane into Vanishing Regions

In this subsection the problem of choosing a convenient partition of the image plane into vanishing regions is addressed. The following criteria are used:

Equal probability

The partition has to be such that the probability $p_j = \mathbb{P}[G \cap V_j \neq \emptyset]$ that a random line G of the image meets a vanishing region V_j is constant for all regions. Without this equiprobability condition, certain vanishing regions would require many more meeting lines to become meaningful than others, that is, they would not be *equally detectable*, which is not desirable.[2]

This equiprobability condition and the results of the previous subsection imply that the size of V_j increases dramatically with its distance from the image, which agrees with the fact that the *localization error* of a vanishing point increases with its distance from the image. Thus, with the equiprobability condition, the localization error of the vanishing points is obtained as a consequence of their detectability.

Angular precision

The size and shape of the vanishing regions should be in accordance with the angular precision of the detected line segments. Because of the discrete character of the digital image, the supporting line of a segment must be considered as a cone with angle $\theta = \arcsin(1/l)$ (see Figure 8.3). If an alignment has a sufficiently precise orientation for its uncertainty cone to be completely cut by a vanishing region, then this intersection event is counted. Otherwise, the event is uncertain and the segment line is not counted as meeting the vanishing region. Thus, the vanishing regions must have a size comparable to the width of the corresponding vanishing cones.

---

[2] For instance, the partition into regions whose projection into the Gaussian sphere has constant area does not necessarily satisfy this equal-probability condition. This was observed in [LMLK94] in the case of uniformly distributed 3-D lines. In this case, lines almost parallel to the image plane become much less probable than lines that are almost orthogonal. Despite the correction proposed in [LMLK94], this still leads to problems in the detection of vanishing points when the perspective effect is very low (distant vanishing points, or lines almost parallel to the image plane), as observed in [Shu99].

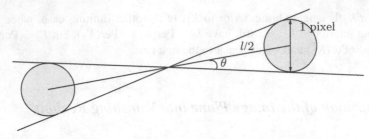

Fig. 8.3 Uncertainty cone of a line segment [Shu99]. The position of the endpoints of a length l segment has a half-pixel error margin. Hence, the supporting line of this segment lies within an uncertainty cone centered at the segment barycenter with angle $d\theta = \arcsin(1/l)$.

Geometric construction

The partition is composed of two families of vanishing regions $\{V_j^{(i)}\}$ and $\{V_j^{(e)}\}$. The first one consists of interior regions entirely contained in the image domain Ω. The second one consists of exterior regions lying outside the image domain. For simplicity, the image domain Ω is approximated by its circumscribed circle with radius R. To meet the angular-precision requirement, all exterior regions V will be portions of sectors of angle 2θ lying between distances d and d' from the image center O. Figure 8.5 illustrates this construction and Figure 8.4 illustrates the trigonometric calculation of the probability that a random line meeting the image domain Ω also meets V. In the case of exterior tiles of angular precision 2θ at distances d and d', this probability becomes

$$
\begin{aligned}
p_e(d,d') &= \frac{L_i - L_e}{\mathrm{Per}(\Omega)} \\
&= \frac{1}{\pi}\left(2\theta + \left[\beta + \frac{1}{\cos\beta} - \tan\beta\right]_{\beta=\arccos(R\cos\theta/d)}^{\beta=\arccos(R\cos\theta/d')}\right) \\
&= \frac{1}{\pi}\left(2\theta + \left[\arccos\left(\frac{1}{q}\right) + q - \sqrt{q^2-1}\right]_{q=\frac{d}{R\cos\theta}}^{q=\frac{d'}{R\cos\theta}}\right). \tag{8.7}
\end{aligned}
$$

Note that it may be occasionally handier to think of p_e as a function of the angles

$$
\beta = \arccos(R\cos(\theta)/d) \quad \text{and} \quad \beta' = \arccos(R\cos(\theta)/d')
$$

instead of the distances d and d' (see Figure 8.4) .

Concerning the interior regions, the disk Ω is simply partitioned into square tiles. The side of each square is chosen to be equal to the side of the exterior tiles closest to the image domain (i.e. $2R\sin\theta$). The perimeter of the interior regions is therefore equal to $8R\sin\theta$, and the probability that a line meets an interior vanishing region is

$$
p_i = \frac{\mathrm{Per}(V)}{\mathrm{Per}(\Omega)} = \frac{4\sin\theta}{\pi}. \tag{8.8}
$$

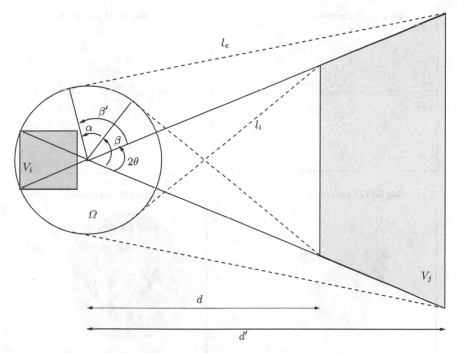

Fig. 8.4 Construction of the exterior vanishing regions V_j, and computation of the corresponding probability $(I_{\cdot i} - I_{\cdot e})/\mathrm{Per}(\Omega)$ that a random line meeting the image domain Ω also meets V_j. To compute the probability p_e of a line meeting an external tile V_j observe that $L_i - L_e$ is composed of two arcs of circle of angle α, the upper and lower sides of V_j of length $q' - q = (d' - d)/\cos\theta$, and two line segments of length $l_i - l_e$. Hence, $L_i - L_e = 2(R\alpha + q' - q + l_i - l_e)$. Observing two right triangles with angles β and β' leads to $l_i = R\tan\beta$ and $l_e = R\tan\beta'$. Finally, since $\alpha + \beta = 2\theta + \beta'$ and $\mathrm{Per}(\Omega) = 2R\pi$, substituting all equations into (8.6) yields (8.7).

This ensures that all interior regions have the same probability and that their size is in accordance with the coarsest angular precision θ of the line segments. Then the values of d and d' are chosen to ensure that all exterior regions have the same probability $p_e = p_i$. One way to do so is to start with the first ring of exterior regions setting $d_1 = R$, and then choose d'_1 by solving the equation $p_e(d_1, d'_1) = p_i$ for d'_1.[3] Then the second ring of exterior tiles is filled by setting $d_2 = d'_1$ and solving the equation $p_e(d_2, d'_2) = p_i$ for d'_2. This process is iterated until $d' \geq d_\infty$, where d_∞ is such that

$$\lim_{d' \to \infty} p_e(d_\infty, d') = p_i. \tag{8.9}$$

To compute this limit, observe that for $d' \to \infty$ we have $\beta' \to \pi/2$ and then $(1/\cos\beta' - \tan\beta') \to 0$; hence,

---

[3] This can be formulated as finding the zero of a convex function of β'_1 with a known derivative. A modified Newton method can be applied that ensures a solution within a given precision on both the x and y axes.

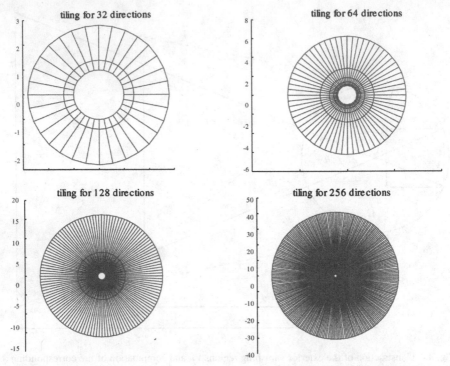

Fig. 8.5 Tilings of the image plane by equal-probability vanishing regions for different angular-precision levels. Only the exterior tiles are shown, except for the last ring of unbounded tiles, whose probability may be smaller. The interior tiles form a regular square partition of the circle with squares of size comparable to the "innermost ring" of exterior tiles. Observe how the size of the tiles increases for the more distant tiles. The axes represent distances relative to the radius of the circular image domain. Note that for higher angular precisions, the image has been zoomed out to allow the visualization of the more distant tiles.

$$\lim_{d' \to \infty} p_e(d, d') = \frac{1}{\pi} \left(2\theta + \frac{\pi}{2} - \beta - \frac{1}{\cos\beta} + \tan\beta \right), \text{ where } \beta = \arccos\left(\frac{R\cos\theta}{d} \right).$$

From the previous equation it can easily be deduced that the value of d_∞ satisfying Equation (8.9) is finite and satisfies

$$4\sin\theta = 2\theta + \frac{\pi}{2} - \beta_\infty - \frac{1}{\cos\beta_\infty} + \tan\beta_\infty, \qquad \text{where} \quad \beta_\infty = \arccos\left(\frac{R\cos\theta}{d_\infty} \right).$$

$$\tag{8.10}$$

Regions in the last ring will then be unbounded, with probability less than p_i. They represent parallel lines in the image plane. Figure 8.5 shows some examples of this partition of the image plane for different precision levels θ.

8.2.4 Final Remarks

This subsection introduces some additional criteria to suppress spurious vanishing points and to eliminate the angular-precision parameter θ.

Multiprecision analysis

The choice of a fixed value for the angular-precision parameter θ requires a compromise between detectability and the localization error of vanishing points. We aim at the highest possible precision level (i.e., the smallest localization error of vanishing points). On the other hand, if the precision level is too fine with respect to the angular precision of the segments, the vanishing region will be hardly detected. The optimal level will approximately match the precision of the segments converging to this vanishing point, and the strategy will be to try to adjust the precision level automatically to this value. Figure 8.6 shows the value of the minimal number of

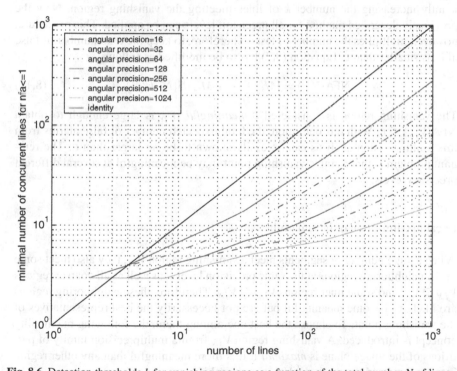

Fig. 8.6 Detection thresholds k for vanishing regions as a function of the total number N of lines detected in the image. These thresholds depend on the joint precision level $\theta = 2^{-s}\pi$ (for $s = 4, 5, \ldots, 10$) of the alignments and regions. N denotes the number of random lines meeting the image domain Ω. The seven curves plot the minimal number $k(N)$ of lines that should meet V_j for V_j to be 1-meaningful for seven precision levels s. Taking (for example) $N = 10$, the plot shows that at high precision (bottom curve, angular precision = 1024) only four lines are necessary to create a vanishing point. Unfortunately, the low precision of the top curves (small angular precisions) is more realistic.

concurrent lines needed for the vanishing region to be 1-meaningful, as a function of the total number N of lines in the image for several angular precisions θ. This figure, or simple calculations using the definition of the NFA, shows that for a total $N = 1000$ lines, about 300 concurrent lines are needed to be meaningful at precision $\theta = \frac{\pi}{16}$, whereas only 15 concurrent lines are enough at precision $\theta = \pi/1024$. However, only 7 concurrent lines would be needed if the total number of lines were $N = 100$. This discussion leads to the procedure described below.

As in the case of alignments, instead of fixing a single angular-precision level, multiple dyadic-precision levels $\theta = 2^{-s}\pi$, for n different values of s in a certain range $[s_1, s_n]$, have to be considered. In the experiments, $s = 4, 5, 6, 7$ proved to be the most useful range, but this can be adjusted to the range of precision levels of the extracted segments. According to the above discussion, at each precision level $\theta_s = 2^{-s}\pi$, only those segments with a precision level no coarser than θ_s are kept. Let N_s denote their number and let M_s denote the number of vanishing regions obtained by the construction described in the previous subsection with angular precision θ_s. Segments with precision coarser than θ_s are not kept, since they would significantly increase N_s (thus increasing the detection threshold k) without significantly increasing the number k of lines meeting the vanishing region. Now the previously described method for all precision levels can be applied. This procedure, however, multiplies the number of tests. Therefore, to keep the total number of false alarms smaller than ε, Equation (8.2) has to be modified as follows:

$$\text{NFA}(V_{j,s}) = (M_{s_1} + \cdots + M_{s_n})\mathcal{B}(N_s, k, p_s). \qquad (8.11)$$

The vanishing region is considered ε-meaningful if k is large enough to obtain $\text{NFA}(V_{j,s}) \leq \varepsilon$. With this definition, the total expected number of false alarms from this multiprecision analysis can be easily shown to be no larger than ε. The remnant problem is that a single vanishing point may be meaningful at several different precision levels.

Local maximization of meaningfulness

When a huge number of segments meet a vanishing region $V_{j,s}$, they also meet some of the neighboring regions at the same precision level s, as well as all coarser regions $V_{j,s'} \supseteq V_{j,s}$ and some finer regions $V_{j,s''} \subseteq V_{j,s}$. Therefore, these neighboring regions are likely to become meaningful but are not necessarily the best representatives of the vanishing point. To choose the best one among them, the following maximality concept is introduced. A vanishing region $V_{j,s}$ from a multiprecision family of partitions of the image plane is *maximal* if it is more meaningful than any other region intersecting it. More precisely, $V_{j,s}$ is maximal if

$$\forall s' \in [s_1, s_n], \forall j' \in \{1, \ldots, M_{s'}\}, \qquad \overline{V_{j',s'}} \cap \overline{V_{j,s}} \neq \emptyset \implies \text{NFA}(V_{j',s'}) \geq \text{NFA}(V_{j,s}),$$
$$(8.12)$$

where $\overline{A}$ denotes the closure of a set A. Note that the condition $\overline{V_{j',s'}} \cap \overline{V_{j,s}} \neq \emptyset$ includes both neighboring regions at the same level, as well as coarser regions containing $V_{j,s}$ and finer regions contained in it.[4]

Exclusion principle for vanishing regions

Figure 8.7 shows all the maximal 1-meaningful vanishing regions that are detected in the photograph of a building. Clearly, the first three correspond to real orientation in the 3-D scene, whereas the other three are an artificial mixture of different orientations. Observe that these mixtures are less meaningful than the original ones because

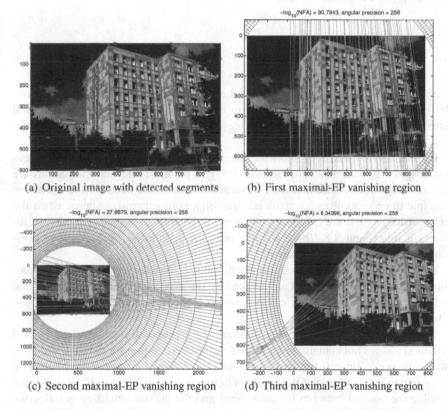

(a) Original image with detected segments   (b) First maximal-EP vanishing region

(c) Second maximal-EP vanishing region   (d) Third maximal-EP vanishing region

Fig. 8.7 Detected line segments for a building image and the only three maximal-EP meaningful vanishing regions that are detected. They correspond to the two horizontal orientations and to one vertical orientation. After this detection, an orthogonality hypothesis between vanishing points could be used to calibrate some camera parameters. For each vanishing region, only the segments that contributed to this region are shown.

---

[4] This condition is used instead of inclusion, because the equal-probability constraint that is used to construct the partition implies that regions at precision level $s+1$ cannot always be completely included in a single region at the coarser precision level s. In this situation, the proposed nonempty intersection-type condition is better suited.

only a small portion of the segments in each direction can participate. Therefore, these artificial vanishing regions can be filtered out by the exclusion principle in a way similar to the one used for segments. Among all maximal meaningful vanishing regions, a contest can be organized, based on the principle that each segment has to choose a single vanishing region that best explains its orientation. More precisely, a segment with supporting line l is assigned to the vanishing region $V_{j,s}$ such that $\text{NFA}(V_{j,s})$ is smallest among all regions $V_{j,s}$ met by l. Then $\text{NFA}(V_{j,s})$ is recomputed for all meaningful segments using Equation (8.11) with the only modification that instead of k, we consider $k' \leq k$, which is the number of lines that not only meet $V_{j,s}$ *but also have been assigned to the vanishing region $V_{j,s}$*. We say that the resulting vanishing regions are maximal for the exclusion principle, and maximal meaningful when their NFA is smaller than ε.

8.3 Experimental Results

The total number N of lines is usually of the order of some hundreds, and the probability parameter $p_s = 4\sin(2^{-s}\pi)/\pi$ can become very small. Thus, the binomial distribution $\mathcal{B}(N, k, p_s)$ can be replaced by its Poisson approximation with parameter $\lambda_s = N p_s$.

Figures 8.7, 8.9, and 8.11 show the results of detecting vanishing points in several images.[5] In man-made environment images, the most relevant orientations are detected without any false alarms. Figure 8.8 illustrates the need for the exclusion principle in order to filter out artificial vanishing points that may appear when the real vanishing points are extremely meaningful. Note that after applying the exclusion principle (Figures 8.7(b) to 8.7(d)), only the main three directions (two horizontal and one vertical) are obtained.

Figure 8.10 illustrates the masking phenomenon. Here the less meaningful directions corresponding to the wall are masked by the many segments in the horizontal and vertical directions, but they can be unmasked. See the figure caption for a more detailed explanation.

Finally, Figure 8.11 shows the limitations of the proposed method when applied to natural images not containing vanishing points (see caption for details). This and other similar experiments further enforce the conclusion of the previous chapters on the importance of addressing the conflicts between gestalts. Indeed, if we were able to solve the conflict between the alignment and the curved boundary gestalts, we would eliminate many wrong line segments and thus further reduce the NFA in the vanishing-point-detection phase.

---

[5] In all of the experiments, $\varepsilon = 1$. A much smaller value could have been used. Indeed, in all of the examples presented here, all real vanishing points have NFA < 0.0001. Furthermore, $\varepsilon = 1$ means that we can expect, on average, one vanishing point in a random segment distribution. This yields a quite high error rate. Indeed, only a few vanishing points are usually found in a digital image.

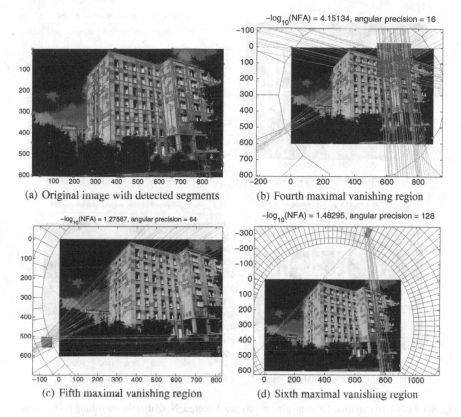

$-\log_{10}(\text{NFA}) = 4.15134$, angular precision = 16

(a) Original image with detected segments

(b) Fourth maximal vanishing region

$-\log_{10}(\text{NFA}) = 1.27587$, angular precision = 64

$-\log_{10}(\text{NFA}) = 1.48295$, angular precision = 128

(c) Fifth maximal vanishing region

(d) Sixth maximal vanishing region

Fig. 8.8 Before applying the exclusion principle, some spurious vanishing regions remain. Note that they arise from mixtures of real vanishing regions, and that they are significantly less meaningful and less precise than the real vanishing regions. Therefore, during the exclusion principle step, most segments vote for the real vanishing region instead of these mixed ones. Thus, after exclusion principle, their NFA increases and they are no longer meaningful.

8.4 Bibliographic Notes

This chapter is based on the work of Almansa et al. [ADV03].

Why are vanishing points needed?

The usefulness of precise measurements of vanishing points, among other geometric primitives, has been demonstrated in many different frameworks [CRZ00, LCZ99, FL01, JR95].

A common situation in architectural environments is to find a set of three orthogonal dominant orientations. If the corresponding vanishing points are detected in the image, they provide three independent constraints on the five internal calibration parameters of the camera [LCZ99]. More importantly, it is very common

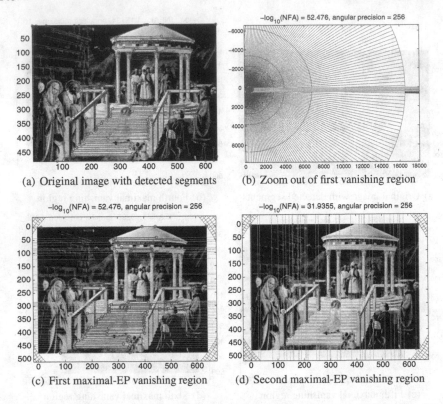

(a) Original image with detected segments   (b) Zoom out of first vanishing region

(c) First maximal-EP vanishing region   (d) Second maximal-EP vanishing region

Fig. 8.9 Only two maximal-EP vanishing points are detected. Note that the vanishing points corresponding to the oblique wall and the staircase are missing. Indeed, both the alignment detection and the vanishing point detection are global, and the less-meaningful segments and vanishing points are masked by the more-meaningful horizontal and vertical orientations (see Figure 8.10).

to have cameras with zero skew and aspect ratio equal to 1 (a natural camera). In this case, the internal parameters are reduced to three, namely the focal length and the 2-D position of the principal point (the orthogonal projection of the focal point into the image plane). Then the camera can be calibrated from a single image of a building – for instance, with the only assumption that the walls and the floor are orthogonal to each other [LCZ99].

Even more impressive is the result described in [CRZ00], where it is shown that (without any knowledge of camera calibration or position) the ratios of lengths between two parallel segments in 3-D can be computed from the lengths of the imaged segments if we know only the vanishing line of a plane containing one of the endpoints of both segments and the vanishing point of the two parallel segments. A typical application of this result is to measure the height of objects standing vertically on the floor relative to a reference object standing vertically on the floor

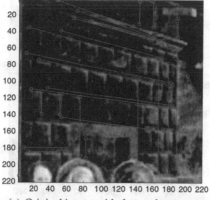

(a) Original image with detected segments

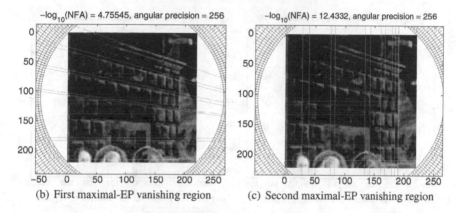

(b) First maximal-EP vanishing region   (c) Second maximal-EP vanishing region

Fig. 8.10 Illustration of the masking phenomenon. When we select the wall subimage in Figure 8.9, more alignments are detected and a new vanishing point that was masked in the global image becomes meaningful. This is due to two cooperating effects. First, the masking phenomenon at the alignment-detection level means that we detect in this subimage more-meaningful segments than in the global image. Second, at the vanishing-point-detection level, the total number of segments is smaller, which means that the required number of concurrent lines for a vanishing region to become meaningful is smaller. A similar result may be obtained by applying a second iteration of the exclusion principle after all segments contributing to the first iteration have been removed.

parallel to them.[6] The only points that are required are two horizontal vanishing points, a vertical vanishing point, and the endpoints of the target and reference segments. Then the computations involve only measuring ratios of lengths of segments defined by these points.

Finally, when using vanishing points in conjunction with other properties such as orthogonality, common segments between planes, and a few length ratio, full 3-D

---

[6] Note that here we used the word "vertical" for clarity and to fix ideas, but actually no orthogonality relationship is required, only 3-D parallelism.

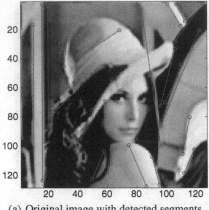

(a) Original image with detected segments

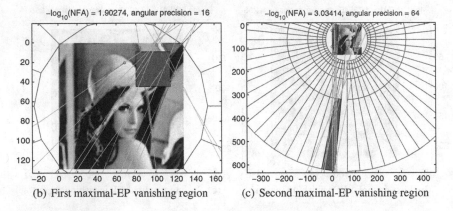

(b) First maximal-EP vanishing region   (c) Second maximal-EP vanishing region

Fig. 8.11 Accidental vanishing points. When applied to images of man-made environments that actually contain vanishing points, the method very rarely detects accidental vanishing points. However, this does happen in natural images in which we do not perceive such vanishing points. Here we show one of the worst such examples that we found in experiments. In this case, the detected vanishing points are probably not perceived because they are made up mostly of segments that are not perceived as straight lines in the first place. Many of these segments would be better explained as meaningful curved boundaries, and therefore will never give rise to vanishing points. Hence, the false alarms in the vanishing-point-detection phase are here to some extent the result of some special kind of false alarms in the alignment detection phase. Further experiments on natural images showed this kind of false alarm of vanishing points (due to some false alarms in line segments that are actually curved boundaries) to be the most prominent one.

reconstructions are possible from a single view, even from Renaissance paintings with carefully respected perspective laws [LCZ99].

When several views are available, vanishing point correspondences can be helpful in determining the epipolar geometry or fundamental matrix [FL01]. Alternatively, if the intrinsic parameters are known, vanishing points can be used to find some information about extrinsic parameters (i.e., the relative position of the two cameras). In [AT00], for instance, vanishing points were used to decouple the relative rotation between the two cameras from the relative translation to which vanishing points are insensitive.

Available methods for vanishing point detection

Since the seminal work of Barnard [Bar83], automated computational methods for vanishing point detection in digital images have been based on some variation of the Hough Transform in a conveniently quantized Gaussian sphere. Several refinements of these techniques followed, but most recent work suggests that this simple technique often leads to spurious vanishing points [Shu99]. To eliminate these false alarms, most authors considered some kind of joint gestalt combining some other property with 3-D parallelism such as coplanarity and equal distance between lines [SZ00] or orthogonality between the three main 3-D directions [LMLK94, Shu99, Rot00]. In addition, knowledge of the intrinsic camera calibration parameters is commonly assumed [LMLK94, AT00] by these methods, or they are designed mostly for omnidirectional images [AT00]. To the best of our knowledge, the question of reliably determining whether an image *actually contains* some vanishing points and its *number* had not been addressed systematically before the work of Almansa et al. [ADV03].

Let us mention [TPG97] for proposing a partition of the plane assigning the same precision to all 3-D orientations. The drawback of the method is that it requires knowledge of the internal camera calibration parameters. Most other works use a partition of the image plane such that the projection of each vanishing region on the Gaussian sphere has a quasi-constant area [Bar83, LMLK94, Shu99, Rot00].

It is quite difficult to build an experimental setup that fairly compares the method described here with previously proposed ones. The reason is that the assumptions are quite different, since the treated problem is not the same. Whereas most previous works [LMLK94, Shu99, Rot00, SZ00] look for joint gestalts that combine 3-D parallelism with some other property, here it was attempted to push the pure partial gestalt of 3-D parallelism to its limits.

An exception is the recent work in [AT00], which relies only on 3-D parallelism and has been shown to produce highly accurate vanishing points. However, it assumes knowledge of the camera calibration parameters and omnidirectional images. The importance of this knowledge is not thoroughly discussed in [AT00] but was crucial in [LMLK94] in order to reduce spurious responses. The work in [AT00] relies on a Hough Transform as in [LMLK94] to determine the number of vanishing points and is therefore prone to the same sensitivity to internal calibration parameters. For this reason, the method proposed in [AT00] can be considered as complementary to the method described in this chapter. It could be used either in the initialization step to determine the number and approximate positions of vanishing points more reliably or as a validation step to reduce the number of false alarms.

8.5 Exercises

8.5.1 Poincaré-Invariant Measure on the Set of Lines

Let $\mathcal{G}$ denote the set of lines in the plane $\mathbb{R}^2$. A line $G \in \mathcal{G}$ is parameterized by its polar coordinate (ρ, θ), which means that it is defined by

$$G = G(\rho, \theta) = \{(x,y) \in \mathbb{R}^2 : x\cos\theta + y\sin\theta = \rho\}.$$

We consider all of the measures μ on $\mathcal{G}$ that are of the form $d\mu = f(\rho, \theta)d\rho\, d\theta$, where $f \geq 0$ is defined and integrable on $\mathbb{R}_+ \times [0, 2\pi)$. Prove that $\mu(T(X)) = \mu(X)$ for all measurable $X \subset \mathcal{G}$ and all translations and rotations T, if and only if f is a constant.

This means that the only translation and rotation invariant measure is, up to a multiplicative constant,

$$dG = d\rho\, d\theta.$$

8.5.2 Perimeter of a Convex Set

Let $K \subset \mathbb{R}^2$ be a closed bounded convex set with nonempty interior $\overset{\circ}{K}$. Assume that the origin 0 belongs to $\overset{\circ}{K}$. The support function $\theta \mapsto p(\theta)$ of K being defined by

$$\forall \theta \in [0, 2\pi], \ p(\theta) = \max_{(x,y) \in K} (x\cos\theta + y\sin\theta) = \max_{(x,y) \in \partial K} (x\cos\theta + y\sin\theta), \quad (8.13)$$

show that the perimeter of K, denoted by $\mathrm{Per}(K)$, is given by the Poincaré formula

$$\mathrm{Per}(K) = \int_0^{2\pi} p(\theta)\, d\theta.$$

Hint: Start with the proof of this formula for a polygon.

8.5.3 Crofton's Formula

We use the notation of the two previous exercises. Let μ denote the measure on the set of lines defined by $d\mu = d\rho\, d\theta$. Let K be a closed bounded convex set with a nonempty interior, and assume that the origin 0 belongs to $\overset{\circ}{K}$. For a set $A \subset \mathbb{R}^2$, we will denote by G_A the set of lines meeting A.

1) Use the result of the previous exercise to prove Crofton's formula:

$$\mu(G_K) = \mu[G : G \cap K \neq \emptyset] = \mathrm{Per}(K).$$

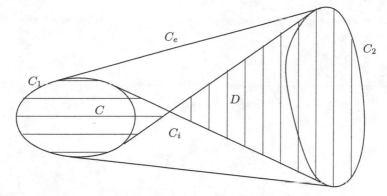

Fig. 8.12 Two convex sets C_1 and C_2, their convex hull C_e, and their "internal envelope" C_i, which is the union of the two convex sets C and D.

2) Let C_1 and C_2 be two closed bounded convex sets such that $C_1 \subseteq C_2$. Show that the measure of all lines meeting both sets is

$$\mu(G_{C_1} \cap G_{C_2}) = \mu[G \cap C_1 \neq \emptyset \text{ and } G \cap C_2 \neq \emptyset] = \text{Per}(C_1).$$

3) Let C_1 and C_2 be two closed bounded convex sets such that $C_1 \cap C_2 = \emptyset$. Let C_e denote the convex hull of C_1 and C_2, and let C_i denote their "internal envelope." Then C_i is the union of the two convex sets C and D (see Figure 8.12).

First, prove that the measure of the set of all lines meeting both sets is

$$\mu(G_{C_1} \cap G_{C_2}) = \mu(G_C \cap G_D).$$

Then prove that

$$\mu(G_C \cap G_D) = \mu(G_C) + \mu(G_D) - \mu(G_{C_e}).$$

Finally, conclude that

$$\mu(G_{C_1} \cap G_{C_2}) = \text{Per}(C_i) - \text{Per}(C_e).$$

4) General case: Let C_1 and C_2 be two closed bounded convex sets. Let C_e denote the convex hull of $C_1 \cup C_2$. Prove that

$$\mu(G_{C_1} \cap G_{C_2}) = \text{Per}(C_1) + \text{Per}(C_2) - \text{Per}(C_e).$$

Chapter 9
Contrasted Boundaries

9.1 Introduction

Among gestalt principles, the color constancy principle is probably the most basic principle and the easiest to simulate. It states that points with the same color (or gray level) that touch each other are automatically grouped. Since by the Shannon principle an image is a continuous function, we can definitely apply this principle to the *level lines* of the image, namely the curves along which the gray level $u(x, y)$ is constant.

Looking at the level lines of several images will convince us that noncontrasted level lines are masked and do not contribute to the geometric perceptual organization. This organization is, instead, conspicuous in contrasted level lines and also leads to an easy visual detection of T- and X-junctions. Thus, we intend to define a *contrasted level line* detector picking out only unmasked level lines. As in the former chapters, this gestalt will be defined *a contrario*. A contrasted level line is a gestalt if and only if it could not happen by chance in a noise image. The next chapter will provide a detailed comparison of this method with earlier and now classical *edge detection* methods such as the active contours method. All that we will say will be restricted to gray-level images but easily extended to color images.

9.2 Level Lines and the Color Constancy Principle

Level lines have several structural properties that make them particularly fit for gestalt analysis. By the Shannon principle, the image is a C^∞ interpolation of the pixels, so that Sard's theorem ensures existence of a finite number of regular level lines at almost all levels. This construction can be made much more explicit if we use a simpler and more local interpolation. All experiments displayed here will show level lines computed by the *bilinear interpolation*. All of the level lines obtained by this particular method have the following properties (for the proof, see Exercise 9.7.1 at the end of this chapter):

- Level lines are closed Jordan curves.
- Two level lines at different levels do not meet.
- Thanks to the Jordan curve theorem, level lines therefore form a set partially ordered by the order $C \leq C'$ if the Jordan curve C is surrounded by C'.
- We call the set of level lines (along with their levels) a *topographic map*; this partially ordered set is a tree.
- The image can be reconstructed from the topographic map.

We call the whole set of level lines at quantized levels, computed by bilinear interpolation, a *digital topographic map of the digital image.* In the topographic map, the orientation of each level line given by the gradient is also kept. Since the images are usually quantized with 256 levels, we will take 256 levels. A representation with more levels would be redundant. The graphic representation will be by far visually more understandable if we only present quantized versions where only levels multiple of 20 or 30 are displayed. Otherwise, since the set of level lines is essentially dense, one only sees a black bunch of curves. Of course, algorithms dealing with the topographic map will keep all levels. Quantization is just a trick to better understand the geometric organization of the topographic map.

Figure 9.2 gives the topographic map of a digitized geometric drawing (Figure 9.1). As usual for visibility, we only show quantized levels at gray-level values multiple of 20. This image is a digitization of a part of Figure 2.10. More intricate examples are given in Figures 9.4, 9.5, 9.7 and 9.8.

Fig. 9.1 A piece of Figure 2.10 page 15 has been digitized to yield a numerical image.

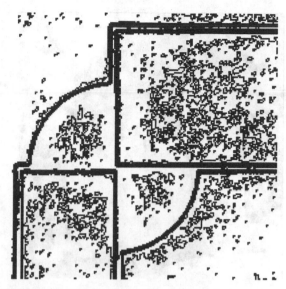

Fig. 9.2 Topographic map of the image in Figure 9.1. For visibility, only 12 levels of the 256 level have been drawn, namely 20, 40, ..., 240. Many level lines seen here are masked, invisible in the original. They have low contrast, are usually short, and show no geometric structure. Thus, they look like level lines in noise and should not be perceived by Helmholtz principle (and they are not). Only the longer and more contrasted level lines can be perceived (see Figure 9.3).

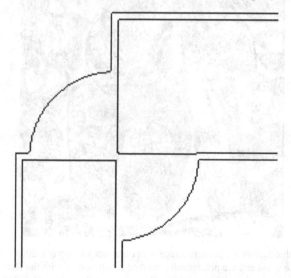

Fig. 9.3 Topographic map of the image in Figure 9.1. Only long level lines of the medium level 128 have been kept. They correspond to visible, unmasked contours in the original. They have a high contrast, are smooth, and contain the essential geometric information of the image. This experiment illustrates the conjecture that image geometry is contained in an adequate choice of image level lines.

Fig. 9.4 A digital image of peppers with 256 gray levels.

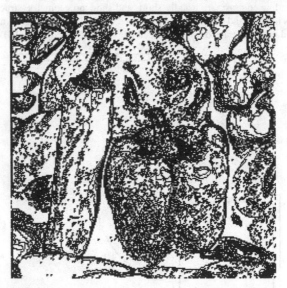

Fig. 9.5 Topographic map of the peppers image with quantization step $q = 20$. Most of the revealed level lines look like noise and are perceptually masked in agreement with the Helmholtz principle. Some level lines correspond to object contours. They are smoother and more contrasted. Since they are unlikely in noise, they can be detected by the Helmholtz principle.

The visualization of the topographic map reveals gestalt principles. The most contrasted contours can be seen as bunches of level lines very close to each other. Notice that most of these contrasted level lines are piecewise regular and therefore doubly conspicuous, first as contrasted and second as smooth. On the other hand, isolated level lines are irregular. Their erratic behavior does not obey any good continuation principle. Consequently, such low-contrasted level lines are usually masked and therefore invisible in the digital image.

These experiments give two guidelines in gestalt analysis of digital images:

1. All useful gestalt information is contained in some level lines.
2. Analysis can be restricted to contrasted or smooth enough level lines. The other ones offer limited perception and could appear in a noise image.

A last illustration of the level line selection problem is given in Figure 9.9. It shows the level lines at three levels of Figure 9.7. Some of them are smooth and straight and can be seen as contours in the original image. The others are long, oscillating, and thoroughly masked in the original.

The *closure principle*, according to which we tend to group all points surrounded by a closed curve as parts of the same object, fits well with the level line representation. Indeed, image level lines are closed curves. When they are contrasted, they surround a perceptual object. The topographic maps of Figures 9.2, 9.3, 9.5 and 9.6 (T-junctions) and of Figures 9.8 and 9.9 (X-junctions) show how level lines arrange themselves in T-junctions or X-junctions and contain a signature of occlusion and transparency phenomena. We refer to Chapter 2 for the definition of these gestalts.

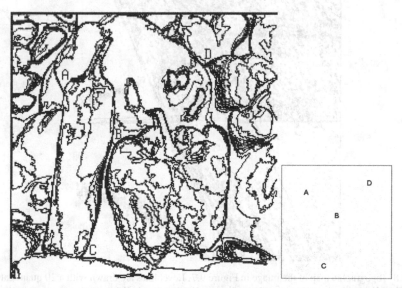

Fig. 9.6 Topographic map of Figure 9.4 with quantization step $q = 30$. The geometric organization of level lines is put in evidence by only conserving those long enough and putting the quantization step to 30 (only 8 equally spaced levels are therefore represented). Notice how T-junctions become apparent (e.g., at points A, B, C and D). These T-junctions, according to Kanizsa, are perceptual cues to the presence of hidden parts. The T-junction at B was not even visible in the original image.

Fig. 9.7 This image has been obtained by digitization on 256 gray levels of a photograph of an envelope partially covered by a transparent slide.

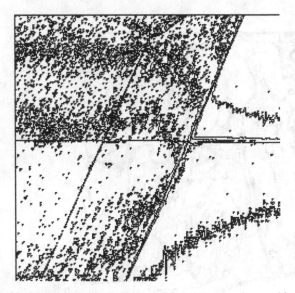

Fig. 9.8 Topographic map of the image in Figure 9.7. Level lines are drawn with a 20 quantization step. Two X-junctions are at sight. According to Kanizsa, such X-junctions are responsible for transparency perception.

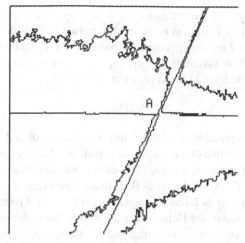

Fig. 9.9 Topographic map of Figure 9.7. Quantization step is $q = 30$ (level lines of levels 30, 60, etc.) We only display long level lines. A single X-junction remains at sight (in A).

9.3 *A Contrario* Definition of Contrasted Boundaries

We call *contrasted boundary* any closed curve, long enough, with strong enough contrast, and fitting well to the geometry of the image. This last requirement is ideally met when the curve is orthogonal to the gradient of the image at each one of its points. In that case, the curve is a level line. We will first define ε-meaningful contrasted boundaries and then maximal meaningful contrasted boundaries. This definition depends on two thresholds (long enough, contrasted enough) that usually remain free in a Computer Vision algorithm. Our aim is to get rid of these parameters and to derive a parameter-free boundary detector. By the Helmholtz principle, we are led to compute the number of false alarms (NFA) of the event "at each point of a length l part of a level line the contrast is larger than μ".

9.3.1 Meaningful Boundaries and Edges

Let u be a $N \times N$ digital image. The level lines are computed at quantized levels $\lambda_1, \ldots, \lambda_k$. The quantization step q is chosen in such a way that the level lines make a dense covering of the image. For example, this quantization step q is 1 when the natural image ranges from 0 to 255. A level line can be computed as a Jordan curve contained in the boundary of a level set with level λ,

$$\chi_\lambda = \{x, \, u(x) \leq \lambda\} \quad \left(\text{or} \quad \chi^\lambda = \{x, \, u(x) \geq \lambda\} \right).$$

The interpolation considered in all experiments below is the bilinear interpolation (see Exercise 9.7.1 at the end of the chapter).

Let L be a level line of the image u. We denote by l its length counted in independent points. In the a-contrario noise model, points at a geodesic distance along the curve larger than 2 are independent. In particular, the gradients at these points are independent random variables. Let $x_1, x_2, \ldots, x_l$ denote the l considered points of L. For $x \in L$, denote by $c(x)$ the contrast at x defined by

$$c(x) = |Du|(x), \tag{9.1}$$

where Du can be computed by a standard finite difference on a 2×2 neighborhood (see Equation (5.2)). Notice that the term "contrast" used here is just the name given to the norm of the gradient at a point (it does not necessarily match its usual definition in perception theories). For $\mu > 0$, consider the event that for all $1 \leq i \leq l$, $c(x_i) \geq \mu$ (i.e. each point of L has a contrast larger than μ). From now on, all computations are performed in the Helmholtz framework. We make all computations as though the image were noise and contrast observations at x_i mutually independent. Thus, the probability of the event is

$$\mathbb{P}[c(x_1) \geq \mu] \cdot \mathbb{P}[c(x_2) \geq \mu] \cdot \; \cdots \; \cdot \mathbb{P}[c(x_l) \geq \mu] = H(\mu)^l, \tag{9.2}$$

where $H(\mu)$ is the probability for a point on any level line to have a contrast larger than μ. An important question here is the choice of $H(\mu)$. Should we consider that $H(\mu)$ is given by an *a priori* probability distribution, or should it be estimated from the image itself (e.g., from the histogram of the gradient norm in the image)? In the case of alignments, by the Helmholtz principle, the orientation at each point of the image was a random, uniformly distributed variable on $[0, 2\pi]$. In the case of contrast, it does not seem sound at all to take a uniformly distributed contrast. The observation of the histogram of the gradient norm of a natural image (see Figure 9.10) shows that most of the points have a "small" contrast (between 0 and 3), and that only a few points are highly contrasted. This is explained by the fact that a natural image contains many flat regions (the so-called "blue sky effect", [HM99]). In the following, we will consider that $H(\mu)$ is given by the image itself, which means that

$$H(\mu) = \frac{1}{M} \#\{x, \; |Du|(x) \geq \mu\}, \tag{9.3}$$

where M is the number of pixels of the image where $Du \neq 0$.

Definition 18 (Number of false alarms). *Let L be a level line with length l, counted in independent points. Let μ be the minimal contrast of the points $x_1, \ldots, x_l$ of L. The number of false alarms of this event is defined by*

$$\mathrm{NFA}(L) = N_{ll} \cdot [H(\mu)]^l, \tag{9.4}$$

where N_{ll} is the number of level lines in the image.

Here, the a-contrario model for the Helmholtz principle is a set of N_{ll} curves, for which at each pixel a random variable is given that follows the repartition function H and such that pixels at a distance larger than 2 are independent. Notice that there

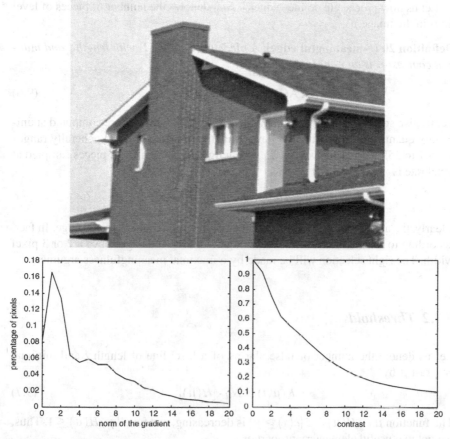

Fig. 9.10 Top: the original image. Bottom-left: histogram of the norm of the gradient. Bottom-right: its repartition function ($\mu \mapsto \mathbb{P}[|Du| \geq \mu]$).

is no noise image here; there is only a noise model. It is generally not possible to build a noise image with a prescribed gradient norm histogram and such that pixels at a distance larger than 2 have independent gradient. Moreover, the level lines in a noise image have a behavior (their number, their length, etc.) that is completely different from the one of level lines in a "natural image" of the same size.

Definition 19 (ε-meaningful boundary). *A level line L with length l and minimal contrast μ is an ε-meaningful boundary if*

$$\text{NFA}(L) = N_{ll} \cdot [H(\mu)]^l \leq \varepsilon. \qquad (9.5)$$

Notice that the number N_{ll} of level lines is provided by the image itself. The above definitions involve two variables: the length l of the level line and its minimal contrast μ. The NFA of an event measures the "meaningfulness" of this event. The smaller it is, the more meaningful the event.

Let us now proceed to define "edges". N_{llp} denotes the number of pieces of level lines in the image.

Definition 20 (ε-meaningful edge). *A piece of level line E with length l and minimal contrast μ is an ε-meaningful edge if*

$$\text{NFA}(E) = N_{llp} \cdot [H(\mu)]^l \le \varepsilon. \tag{9.6}$$

Let us give some details on the computation of N_{llp}. Level lines are computed at uniformly quantized levels. The gray-level quantization step is 1 and generally ranges from 1 to 255. For each level line L_i with length l_i, its number of pieces sampled at pixel rate is

$$N_{llp} = \sum_i \frac{l_i(l_i - 1)}{2}.$$

Clearly, the aim so far is *detection* and not optimization of the detected edge. In fact, according to Shannon conditions, the image is blurry and edges have a 2 or 3 pixel width. The right selection will be made by a maximal meaningfulness argument.

9.3.2 Thresholds

Let us denote the number of false alarms of a level line of length l and minimal contrast μ by

$$F(\mu,l) = N_{ll} \cdot [H(\mu)]^l. \tag{9.7}$$

The function $\mu \mapsto H(\mu) = \mathbb{P}[c(x) \ge \mu]$ is decreasing, and for all μ, $H(\mu) \le 1$. Thus, we get two useful elementary properties.

Lemma 6 *If $l \le l'$, then $F(\mu,l) \ge F(\mu,l')$. Thus, if two level lines have the same minimal contrast the more meaningful one is the longer one. If $\mu \le \mu'$, then $F(\mu,l) \ge F(\mu',l)$; that is, if two level lines have the same length, the more meaningful is the one with higher contrast.*

When the contrast μ is fixed, the minimal length $l_{min}(\mu)$ of an ε-meaningful level line with contrast μ is

$$l_{min}(\mu) = \frac{\log \varepsilon - \log N_{ll}}{\log H(\mu)}. \tag{9.8}$$

Conversely, if the length l is fixed, the minimal ε-meaningfulness contrast $\mu_{min}(l)$ is

$$\mu_{min}(l) = H^{-1}\left([\varepsilon/N_{ll}]^{1/l}\right). \tag{9.9}$$

In practice, these thresholds allow very short contrasted and long level lines with low contrast.

9.3.3 Maximality

In this subsection we address the usual multiple detection problem and try to solve it by a maximal meaningfulness property permitting a selection. Let us start with boundaries. If a level line is meaningful, the nearest level lines are likely to be meaningful too. A natural relation between closed level lines is given by their inclusion [Mon00]. If C and C' are two different closed level lines, then C and C' cannot intersect. Let D and D' denote the bounded domains surrounded by C and C'. Then either $D \cap D' = \emptyset$ or $(D \subset D'$ or $D' \subset D)$. Thus, any set of level lines has an inclusion tree structure. Consider the subtree of ε-meaningful level lines satisfying $F(\mu, l) \leq \varepsilon$. On this subtree, one can, following Monasse define *monotone branches*–that is sequences of level curves C_i, $i \in [1, k]$, such that the following hold:

- For $i \geq 2$, C_i is the unique son of C_{i-1} in the tree.
- The gray levels of C_i are either decreasing from 1 to k, or increasing from 1 to k.
- The branch is maximal (not contained in a longer one).

Many such monotone branches of meaningful curves can be seen in the experiments as bunches of well-contrasted level lines along the image boundaries.

Definition 21. *A level curve of a branch is called an optimal boundary if its false alarms number $F(\mu, l)$ is minimal in the branch. We call the optimal boundary map of an image the set of its optimal boundaries.*

The optimal boundary map will be compared in the experiments with classical edge detectors or segmentation algorithms.

We now address the problem of finding optimal edges among the detected ones. We will not be able to proceed as we did for the boundaries. Although the pieces of level lines inherit an inclusion structure from level lines, we cannot compare two of them belonging to different level curves. Indeed, they can have different positions and lengths. We can, instead, compare two edges belonging to the same level curve. The main aim is to define on each curve a set of disjoint maximally detectable edges. Let us denote by $\text{NFA}(E) = F(\mu, l)$ the false alarm number of a given edge E with a minimal gradient norm μ and length l.

Definition 22 (Maximal meaningful edge). *An edge E is maximal meaningful if for any other edge E' on the same level curve such that $E \subset E'$ (resp. $E' \subset E$) one has $\text{NFA}(E') > \text{NFA}(E)$ (resp. $\text{NFA}(E') \geq \text{NFA}(E)$).*

This definition follows exactly the definitions of maximality already given in the case of alignments (Chapter 6) and in the case of histogram modes (Chapter 7).

Proposition 28 *Two maximal edges cannot meet.*

Proof — Let E and E' be two maximal distinct and nondisjoint meaningful edges in a given level curve and μ and μ' be the respective minima of gradient of the image on E and E'. Assume, for example, that $\mu' \leq \mu$. Then $E \cup E'$ has the same minimum as E' but is longer. By Lemma 6, $F(\mu', l + l') < F(\mu', l')$, which implies that $E \cup E'$ has a smaller NFA than E'. Thus, E' is not maximal. $\square$

9.4 Experiments

• **INRIA desk image (Figure 9.11)**
In this experiment, the meaningful boundary method is compared with two other boundary detection methods: the Mumford-Shah image segmentation and the Canny-Deriche edge detector.

In the simplified Mumford-Shah model [MS85], an image u defined on a domain D is approximated by a piecewise constant image v minimizing the functional

$$E(v) = \int_D |v - u|^2 + \lambda \, \text{length}(K(v)),$$

where $\text{length}(K(v))$ is the length of the discontinuity set of v and λ is a user parameter. This energy is a balance between a fidelity term (the approximation error in L^2 norm) and a regularity term (the total length of the boundaries). The result v, called a segmentation of u, depends on the scale parameter λ. As shown in Figure 9.11, the Mumford-Shah model generally produces reasonable boundaries, except in smooth zones where spurious boundaries often appear (see the front side of the desk for example). This is easily explained: The *a priori* model is that the image is piecewise flat with boundaries as short as possible. The image does not fit exactly the model. The desk in the image is smooth but not flat. The detected wrong boundary decreases the functional by dividing the desk into flat regions. The same phenomenon occurs in the sky of Figure 9.12.

The Canny-Deriche filter [Can86, Der87] is an optimization of Canny's well-known edge detector, roughly consisting in the detection of maxima of the gradient norm in the direction of the gradient. Notice that in contrast with the Mumford-Shah model and with the model of meaningful edges or boundaries, it does not produce a set of boundaries (i.e., one-dimensional structures) but a discrete set of points that still are to be connected. This edge set depends on two parameters: the image previous blurring parameter and a threshold on the norm of the gradient that selects candidates for edge points. As we can see in Figure 9.11, the result is very dependent on this threshold. Notice that many Canny edges are found in flat regions of the image where no perceptual boundary is present. If the threshold is increased as shown on the right, the detected edges look perceptually more correct but are broken.

• **Cheetah image (Figure 9.12)**
This experiment compares the meaningful edge detector with the Mumford-Shah model. The Mumford-Shah model follows an economy principle and detects less boundaries. Yet it finds spurious boundaries in the sky due to the inadequacy of the piecewise constant model. In the experiment, we display all meaningful boundaries, not just the optimal ones.

• **DNA image (Figures 9.13 and 9.14)**
This experiment illustrates the concept of optimal boundaries. Each spot in the image produces several parallel meaningful boundaries due to its important blur.

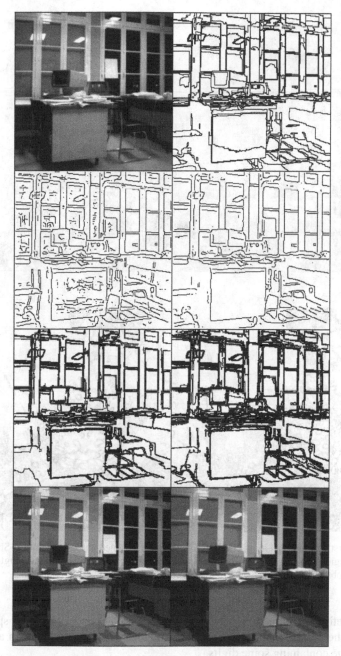

Fig. 9.11 First row: left: original image; right: boundaries obtained with the Mumford-Shah model (1000 regions). Second row: edges obtained with Canny-Deriche edge detector for two different threshold values (2 and 15). Third row: meaningful edges (left) and meaningful boundaries (right) obtained with $\varepsilon = 1$. Fourth row: reconstruction with the Mumford-Shah model (left) and with the set of meaningful boundaries (right). This last reconstruction is easily performed by the following algorithm: Attribute to each pixel x the level of the smallest (for inclusion) meaningful level line surrounding x (see [Mon00]).

Fig. 9.12 First row: original image (left) and boundaries obtained with the Mumford-Shah model with 1000 regions (right). Second row: maximal meaningful edges (left) and meaningful boundaries (right) obtained with $\varepsilon = 1$.

With the definition of maximality, exactly one optimal boundary for each spot is selected. In the second experiment, we compute meaningful boundaries on a subpart of the image containing some digits.

● **Segments image (Figure 9.15)**
As in the DNA experiment, the optimal boundaries correspond to exactly one boundary per object (here, hand-drawn segments). In particular, the number of obtained boundaries (21) counts exactly the number of segments.

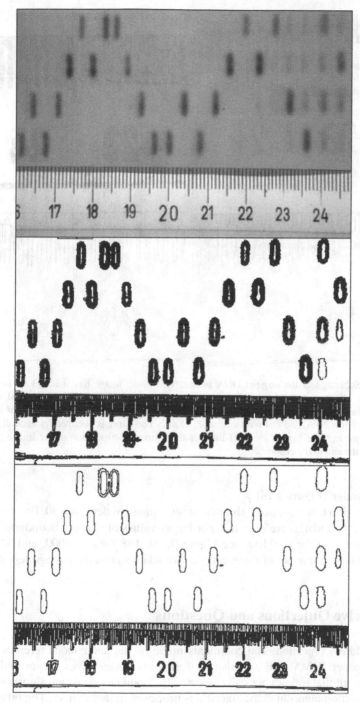

Fig. 9.13 From top to bottom: 1. original image; 2. all meaningful boundaries; 3. optimal meaningful boundaries. The oval blurry spots are surrounded by many meaningful boundaries but by exactly one optimal boundary.

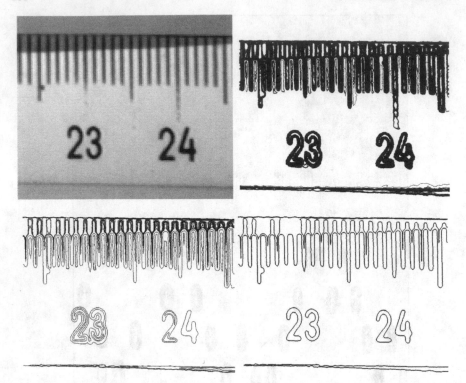

Fig. 9.14 Subimage of the original DNA image. This small image has size 133×114 pixels. We display all meaningful boundaries ($\varepsilon = 1$), optimal meaningful boundaries, and optimal ε-meaningful boundaries for $\varepsilon = 10^{-25}$. This experiment shows that the boundaries of the digits have very low NFA, and that with a small ε, exactly one curve is obtained for each digit. The curves shown in these figures are level lines of the bilinear interpolation of the image (see the exercise at the end of the chapter).

● **Noise image (Figure 9.16)**

This image is a Gaussian noise simulation with standard deviation 40. For $\varepsilon = 1$ and $\varepsilon = 10$, no boundaries are detected. For larger values of ε, some boundaries begin to be detected: 7 for $\varepsilon = 100$ (see Figure 9.16), 148 for $\varepsilon = 1000$, and 3440 for $\varepsilon = 10,000$. The number of ε-meaningful boundaries is on the average less than ε.

9.5 Twelve Objections and Questions

Let us address objections and comments made by the anonymous referees of the original paper [DMM01b] and also by José-Luis Lisani, Yves Meyer and Alain Trouvé. In all that follows, we call respectively "boundary detection algorithm" and "edge detection algorithm" the algorithms proposed in this chapter. The other edge or boundary detection algorithms put into the discussion will be called by their author's names (Mumford-Shah, Canny).

Fig. 9.15 Top-left: original image. Top-right: all the level lines with gray-level step equal to 5. Bottom-left: all meaningful boundaries. Due to the image blur several ones surround each pencil stroke. Bottom-right: optimal meaningful boundaries. Only one is left for each pencil stroke.

Objection 1. The blue sky effect

If a significant part of a natural image happens to be very flat because of some blue sky effect, then **most** level lines of the image will be detected as meaningful. If, for example, one-tenth of the image is a black flat region, then the gradient histogram has a huge peak near zero. Thus, all gradients slightly above this peak will have a probability $\frac{9}{10}$, a number significantly smaller than 1. As a consequence, all level lines long enough (with length larger than, say, 30 pixels) will be meaningful. In practice, this means that the image will be plagued with detected level lines with a small contrast.

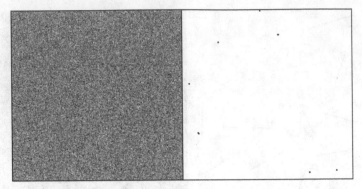

Fig. 9.16 Left: an image of a Gaussian noise with standard deviation 40. Right: the meaningful boundaries found for $\varepsilon = 100$ (no boundaries are found for $\varepsilon = 1$). This agrees with the Helhmholtz principle that, on average, no detection should occur in noise.

Answer 1. If the image has a wide blue sky, then most level lines of the ground are meaningful because any strong deviation from zero becomes meaningful. This effect can be checked in Figure 9.12: The structured and contrasted ground has many detected boundaries (and the sky has none). This outcome can be interpreted in the following way. When a flat region is present in the image, it gives, via the gradient histogram, an indirect noise estimate. Every gradient that is above the noise gradient of the flat region is meaningful.

Question 2. Why not use global known statistics of images?
Instead of using the histogram of the norm of the gradient in the image, one could use some known global statistics about the gradient in natural images (as it is for instance done in the paper of Simoncelli and Adelson [SA96] for noise removal or in the paper of Heiler and Schnörr [HS05] for image segmentation).

Answer 2. In the definition of meaningful boundaries, we can, of course, replace the term $H(\mu)$ by an estimate (from a large database of natural images) of the probability to have a gradient norm larger than μ. Then the detection thresholds will no longer depend on each image. However, this is rather a drawback: If the image is almost flat everywhere and contains just an object with very low contrast, then nothing will be meaningful according to global statistics, whereas the boundary of the object will become meaningful according to the statistics of the image. This corresponds also to our perception: When looking at an image, we adapt our detection thresholds to this image.

Objection 3. Dependence on windows
The detection of a given edge depends on the window containing the edge on which the algorithm is applied?

Answer 3. Yes, the algorithm is global and is affected by reframing the image. For example more edges will be detected in a low-contrasted window (see Figure 9.17).

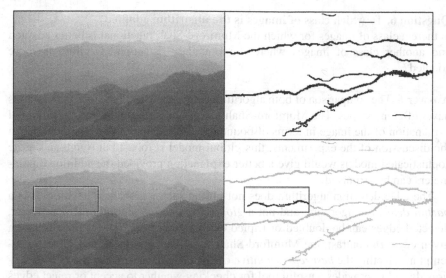

Fig. 9.17 Edge detection is context dependent. First row left: original image (Chinese landscape); right: maximal meaningful edges for $\varepsilon = 1$. Second row: the same algorithm, but run on a sub-window (drawn on the left image); right: the result in black, with in light gray the edges that were detected in the full image.

Question 4. How do you compute edges with multiple windows?
With such a theory, one can apply the detection algorithm on many overlapping windows of the image and in that way let edges multiply!

Answer 4. If a window is too small, no edge at all will be detected in it. If the algorithm is applied to say 100 windows, the number of tests is increased. Thus, the value of ε in each window must decrease accordingly. One can take on each one $\varepsilon = 1/100$. Then the global number of false alarms over all windows remains equal to 1. Thus, a multiwindows version of the algorithm is feasible and recommended. Indeed, psychophysics and neurophysiology both indicate a spatially local treatment of the retina information.

Objection 5. Synthetic images where everything is meaningful
If an image has no noise at all (for instance, a synthetic image), all boundaries contain relevant information. Won't the algorithm detect them all?

Answer 5. Right. If a synthetic binary image is made, for example, of a black square with white background, then all gradients are zero except on the square's boundary. The gradient histogram has a single value 255. (Remember that zero values are excluded from the gradient histogram.) Thus, $H(255) = 1$, which means that no line is meaningful. Thus the square's boundary will not be detected, which is a bit paradoxical! The addition of a tiny noise or of a slight blur would, of course, allow the detection of this square's boundary. This means that synthetic piecewise constant images fall out of the range of the detection algorithm.

Question 6. To which class of images is the algorithm adapted?
Is there a class of images for which the Mumford-Shah functional is better adapted and another class of images where the boundary detection algorithm is more adapted?

Answer 6. The comparison of both algorithms may be misleading since the methods have different scopes. The Mumford-Shah algorithm aims at a global and minimal explanation of the image in terms of boundaries and regions. As we pointed out in the discussion of the experiments, this global model is robust but rough and more sophisticated models would give a better explanation provided the additional parameters can be estimated.

The edge detection algorithm does not aim at such a global explanation. It is a *partial detection* algorithm, and not a *global explanation* algorithm. In particular, detected edges can be doubled or tripled or more since many level lines follow a given edge. In contrast, the Mumford-Shah functional and the Canny detector attempt at selecting the *best representative* of each edge. Conversely, the edge detection algorithm provides a useful tool for checking whether to accept or reject edges suggested by any other algorithm.

Objection 7. The algorithm depends on the quantization step
The algorithm depends on the quantization step q. When q tends to zero, there will be more and more level lines. Thus, N_{ll} and N_{llp} (numbers of level lines and pieces of level lines respectively) will blow up. Thus, there are fewer and fewer detections when q increases and eventually none!

Answer 7. Right again. The numbers N_{ll} and N_{llp} stand for the number of trial tests on the image. When the number of tests tends to infinity, the number of false alarms of Definition 18 also tends to infinity. As we mentioned, q must be small enough to be sure that all edges contain at least one level line. Since the quantization noise is 1 and the standard deviation of noise never goes below 1 or 2, it is not possible to find any edge with contrast smaller than 2. Thus, $q = 1$ is enough if we want to miss no edge.

Question 8. Accuracy of the edges depends on the quantization step?
If q is not very small, accuracy in edge position is lost. Indeed, the quantized levels do not coincide with the optimal level of the edge. A Canny edge detector is more accurate.

Answer 8. Right again. The Canny edge detector performs two tasks in one: detecting and optimizing the edge's position with subpixel accuracy, depending on the interpolation method. The spatial accuracy of the proposed edge detector is roughly $q/\min |Du|$, where the minimum is computed on the detected edge. If, as indicated, $q = 1$ the accuracy is proportional to the inverse of the gradient, as in most edge detectors. This fact will be illustrated in the next chapter, where the boundary detector is compared to the active contour method.

Objection 9. Edges are not level lines
The claim is that every edge coincides with some level line. This is simply not true!

Answer 9. If an edge has contrast kq, where q is the quantization step (usually equal to 1), then k level lines coincide with the edge locally. One can certainly construct long edges whose contrast is everywhere k but whose average level varies in such a way that no level line fully coincides with the edge. However, long pieces of level lines coincide partially with it. Thus, detection of this edge by the detection algorithm is still possible, but it will be detected as a union of several more local edges.

Objection 10. Values of the gradient on the level lines are not independent
The chosen test set is the set of all level lines. The claim is that the gradient amplitudes at two different points of every edge are independent. In most images this is not true.

Answer 10. The independence assumption is made to apply the Helmholtz principle according to which every large deviation from randomness is perceptible. The independence assumption is not intended to model the image. It is an a-contrario assumption against which the gestalts are detected.

Objection 11. A minimal description length model would do the job as well
A Minimum Description Length model (MDL) can contain very wide classes of models for which parameters will be estimated by the MDL principle of shortest description in a fixed language. This fixed language can be the language of gestalt theory, which explains the image in terms of lines, curves, edges, regions, and so forth. Then the existence and nonexistence of a given gestalt would derive from the MDL description: A "detectable" edge would be an edge taken up by the minimal description. (See the work of Leclerc [Lec89] for such a MDL segmentation algorithm.)

Answer 11. A MDL model is all-encompassing in nature. Until it has been constructed, we cannot make any comparison. In a MDL model, the thresholds on edges would depend on all other gestalts. Thus, we would be in the same situation as with the Mumford-Shah model: We have seen that a slight error on the region model leads to false detections for edges. The proposed method is less ambitious. It is a *partial gestalt* detection algorithm, which does not require the establishment of a previous global model.

Objection 12. How many parameters in the method?
We have seen in the course of the discussion no less than three parameters coming out: the choice of the windows, the choice of q, and, finally, the choice of ε.

Answer 12. Always fix $\varepsilon = 1$. Indeed, the dependence of the detection threshold k_{min} on ε is a log-dependence. The q dependence of k_{min} also is a log-dependence since the number of level lines varies roughly linearly as a function of q and the same argument applies to the number of windows.

9.6 Bibliographic Notes

In this chapter we have followed [dFM99] and [FMM98] for the gestalt analysis and [CCM96] where a level-line-based T-junction and X-junction detector is analyzed. The boundary detection algorithm based on the Helmholtz principle is due to [DMM01b]. Extension of the algorithm and further mathematical analysis can be found in an extensive paper by Cao, Musé and Sur [CMS05]. These authors define Helmholtz boundary detectors based on both contrast and smoothness. One of their striking experimental results is that detectors based on level line smoothness give results equivalent to contrast detectors.

In [AADLTT06], Abraham, Abraham, Desolneux and Li-Thiao-Té have defined significant edges in a different framework, but the idea is the same as meaningful edges. They define an edge point as a zero-crossing of the Laplacian, and then significant edge points are the ones that have a gradient above a threshold computed from the noise model. Such significant edge points have a probability less than ε to appear in pure noise images.

All of this chapter has been restricted to gray-level images. As already mentioned, the geometric perception is essentially based on brightness contrast. A discussion of the (weak) role of color in geometry perception can be found in [CCM02].

The fast computation of the topographic map of a digital image and its easy manipulation raise topological and algorithmic concerns that are elegantly solved in [Mon00], [MG00], and [LMR01].

One of the main properties of the topographic map of an image is its invariance by any change in contrast of the image: The set of all level sets $\{\chi_\lambda\}_{\lambda \in \mathbb{R}}$ of an image u remain the same if the image is changed into $g(u)$, where g is an increasing function. The use of level sets for image analysis is one of the main topics of the Mathematical Morphology school (see e.g., [Ser82] , [Ser88] and references therein).

The search for image boundaries (i.e., closed edges) started very early in Computer Vision [Zuc76, Pav86]. We have compared meaningful edges and boundaries with the results of the Canny edge detector [Can86] and with the results of the Mumford-Shah functional for segmentation [MS85]. In the next chapter, we will discuss the link between meaningful boundaries and the "snake" or "active contour" theory and give many more references.

Most boundary detection algorithms like the snake method [KWT87] introduce regularity parameters [MS94]. The requirement that boundaries must be long enough cannot be skipped. The risk is to detect boundaries at all high gradient points as happens with the classical Canny edge detector [Can86]. It is well acknowledged that edges, whatever their definition might be, are as orthogonal as possible to the gradient [Can86, Dav75, DH73, Mar72, RT71]. This led us to define level lines are the effective candidates for edges.

9.7 Exercise

A digital image u is defined as a finite array of gray-level values $u(i,j)$, for $1 \leq i$, $j \leq N$. The values $u(i,j)$ are generally quantized on integer values, from 0 (black) to 255 (white). The typical values for the size N of the image are 256, 512, and 1024. According to the way digital images are generated, the values $u(i,j)$ must be thought of as the samples of an underlying smooth image $u(x,y)$, $(x,y) \in [1,N]^2$. One can use different interpolation methods to compute $u(x,y)$. The best one is the interpolation given by Shannon's theorem, which computes u as the unique trigonometric polynomial function of degree N with the given values $u(i,j)$ at the points (i,j). This method is, however, highly nonlocal and does not provide simple computation and description of the level lines of u. Thus, we will use a much simpler method, namely the *bilinear interpolation*, which computes u as a continuous function.

9.7.1 The Bilinear Interpolation of an Image

1) Consider four neighboring pixels with respective centers (i,j), $(i+1,j)$, $(i+1, j+1)$, and $(i,j+1)$ together with their respective integer gray-level value denoted by a, b, c and d. Let $C_{i,j}$ denote the square whose vertices are the centers of the four pixels. Its center is the point $(i+\frac{1}{2}, j+\frac{1}{2})$. It has side length 1 and is called a "dual pixel."

Show that there exists a unique 4-tuple $(\alpha, \beta, \gamma, \delta)$ such that the function v defined by $v(x,y) = \alpha xy + \beta x + \gamma y + \delta$ has the same value as u at the four vertices of $C_{i,j}$.

Show that the function v thus piecewise defined is continuous on $[1,N]^2$.

2) Let λ be a real number between 0 and 255 and let us consider the set

$$L_\lambda = \{(x,y),\ v(x,y) = \lambda\}.$$

Show that for each dual pixel $C_{i,j}$, the set $L_\lambda \cap C_{i,j}$, when it is nonempty, is one of following:

1. the dual pixel $C_{i,j}$;
2. a piece of hyperbola connecting one side of the square to another side;
3. two disjoint pieces of hyperbola both connecting one side of the square to another side;
4. a line segment connecting one side of the square to another side (give the formula for v in this case);
5. two line segments that are perpendicular and parallel to the sides of the square (give the formula for v in this case).

Notice that the first case only occurs when the four values of u at the vertices of the square are equal. The fifth case only occurs when $Dv = 0$ at a point inside the square and when the meeting point of the two line segments is a saddle point for v.

3) We assume here that λ is not an integer. Using the previous question, show that the set L_λ is the union of a set of piecewise smooth curves that can meet only at a saddle point. Moreover, show that these curves are either closed or open and connecting two image boundary points.

4) The curves that are open are completed into closed curves using the following rule: The two end points belonging to the boundary of the image are connected by the shortest of the two pieces of boundary of the image they define. If these two pieces have the same length, choose the one whose position (center of gravity) is topmost and if still equal, rightmost. The obtained closed curves L_λ, for $\lambda \notin \mathbb{N}$, are called the "level lines" of the image, and their set is called the topographic map of the image.

Show that two different level lines cannot meet. A partial order on the set of level lines is defined by the relation $C \leq C'$ if C' surrounds C (which means that the bounded connected component of $\mathbb{R}^2 \setminus C'$ contains C). Show that the set of level lines has a tree structure with respect to this partial order.

5) Draw the topographic map of the following image ($N = 4$):

$$u = \begin{pmatrix} 3\,3\,3\,5 \\ 3\,4\,4\,3 \\ 2\,5\,0\,1 \\ 1\,0\,2\,3 \end{pmatrix}.$$

6) Compute the gradient $Dv(i + \frac{1}{2}, j + \frac{1}{2})$ at the center of a dual pixel as a function of the values at the four vertices of the square $C_{i,j}$ (see also Chapter 5).

Chapter 10
Variational or Meaningful Boundaries?

10.1 Introduction

This chapter contains a brief review of the theory of edge detection and its nonlocal version – the "snake" or "active contour" theory. The snakes are curves drawn in the image and minimizing a contrast and smoothness energy. Our discussion concludes that a very recent class of snake models proposed by Kimmel and Bruckstein [KB01] [KB02] is optimal. This snake energy is

$$F(\gamma) = \frac{1}{L(\gamma)} \int_0^{L(\gamma)} g\left(\frac{\partial u}{\partial n}(\gamma(s))\right) ds, \tag{10.1}$$

where $\gamma(s)$ is a curve drawn in the image parameterized by length. $L(\gamma)$ is the length of the curve,

$$\frac{\partial u}{\partial n}(\gamma(s)) = Du(\gamma(s)) \cdot \gamma'(s)^{\perp}$$

is the derivative of u at $\gamma(s)$ in the direction normal to the curve, and $g > 0$ is an even contrast function. What can be the best form for the contrast function g? We will prove that a very particular form for g is optimal. Experiments will also demonstrate that the resulting optimal snakes simply coincide with the well-contrasted level lines of the image. Indeed, it will be proven that all meaningful level lines of the image *hardly move* by the snake optimization process. Conversely, evolving curves by their energy brings them back to well-contrasted level lines. Thus, the analysis cross-validates both boundary detection models.

10.2 The "Snakes" Models

The question of how to compute salient image boundaries or edges is unfortunately not outdated. In the past 30 years, many methods have been proposed and none has become a standard. All agree that an edge can be defined as a curve across which the image is contrasted. A first local view of the question would be that edges are made of points where the image gradient is high and maximal in some sense. Hildreth and Marr [MH80] proposed to define the boundaries in a gray-level image $u(x,y)$ as curves across which the Laplacian

$$\Delta u(x,y) = u_{xx}(x,y) + u_{yy}(x,y)$$

changes sign. This definition was based on the remark that for a 1-D function $u(x)$, points with the highest gradient satisfy $u''(x) = 0$. In the same direction, Haralick [Har84] defined "edge points" as the points where the magnitude of the gradient $|Du|$ attains a maximal value along gradient lines. An easy computation shows that such points (x,y) satisfy $D^2u(Du,Du)(x,y) = 0$. We use the notations

$$D^2u = \begin{pmatrix} u_{xx} & u_{xy} \\ u_{yx} & u_{yy} \end{pmatrix}, \quad Du = \begin{pmatrix} u_x \\ u_y \end{pmatrix},$$

and if $A = \begin{pmatrix} a & b \\ b & c \end{pmatrix}$ and $\xi = \begin{pmatrix} v \\ w \end{pmatrix}$, we set $A(\xi,\xi) = \xi^T A \xi = av^2 + 2bvw + cw^2$.

In the following, we will talk about Hildreth-Marr's and Haralick's edge points. The Haralick's edge points computation was proved by Canny [Can86] to be optimal for "step edges" under the Gaussian noise assumption. In other words, it is optimal when the image is a Heavyside function plus a Gaussian white noise.

Edge points have to be connected to form curves. This is why edge detection methods have evolved towards boundary detection methods, *i.e.* methods which directly deliver curves in the image along which the gradient is, in some sense, highest. There is agreement about two criteria. An edge is a smooth curve with high image contrast. The smoothness requirement is relative though, since visual objects can be ragged or have corners.

The smooth $\gamma(s)$ we will consider are one-to-one maps from some real interval $[0,L(\gamma)]$ into the image plane. They are parameterized by length so that

$$|\gamma'(s)| = 1$$

and $L(\gamma)$ is the length of γ. Set $v^\perp = (-y,x)$, a vector orthogonal to $v = (x,y)$. The unit tangent vector to the curve is $\gamma'(s)$ and

$$\mathbf{n}(s) = \gamma'(s)^\perp,$$

a vector normal to the curve γ. We finally consider

$$\gamma''(s) = \text{Curv}(s), \tag{10.2}$$

which is a vector normal to the curve whose magnitude is proportional to its curvature. If $\text{Curv}(s)$ is defined, the curve behaves locally as a circle with radius $1/|\text{Curv}(s)|$. Let $g : \mathbb{R}^+ \to \mathbb{R}^+$ be any even decreasing function. By the contrast requirement, an "edge" is a curve $\gamma(s)$ such that

$$\int_{[0,L(\gamma)]} g(|Du(\gamma(s))|)\,ds$$

is small. By the smoothness requirement,

$$\int_{[0,L(\gamma)]} (1 + |\text{Curv}(s)|) \, ds$$

is also small. Both requirements can be combined in a single energy functional. In the Kass-Witkin-Terzopoulos original model [KWT87], snakes are therefore defined as local minima of the sum of both terms,

$$\int_0^{L(\gamma)} g(|Du(\gamma(s))|) \, ds + C \int_0^{L(\gamma)} (a + |\text{Curv}(s)|) \, ds. \tag{10.3}$$

A typical form for g is $g(s) = 1/(1 + s^2)$. This model has been generally abandoned for the derived "geodesic snakes" due to Caselles, Kimmel, and Sapiro [CKS97]. This model proposed a geometric form of the snakes functional, where the minimum of

$$\int_0^{L(\gamma)} g(|Du(\gamma(s))|) \, ds \tag{10.4}$$

is sought. Since $g > 0$, this functional looks for a compromise between length and contrast. It yields a well-contrasted curve and it can be proven that minima have bounded curvature. Notice, however, that the minimization process tends to decrease the length and forces the snake to shrink, which is not exactly what is desired!

This drawback may explain why Fua and Leclerc [FL90] proposed to minimize the average functional

$$\frac{1}{L(\gamma)} \int_0^{L(\gamma)} g(|Du(\gamma(s))|) \, ds. \tag{10.5}$$

Here, again, g is decreasing. Minimizing this functional amounts to finding a curve in the image with maximal average contrast. One of the main advances in the last two models is the reduction of the number of model parameters to a single one: the contrast function g. The Fua-Leclerc model focuses on contrast only and is therefore apparently no more a hybrid combination of contrast and smoothness requirements.

Both models are in one aspect less accurate than edge detectors defined at the beginning of the seventies. The Montanari [Mon71] and Martelli [Mar72] original boundary detection models used as contrast indicator, instead of $|Du(\gamma(s))|$, a discrete version of

$$u_n(s) = \frac{\partial u}{\partial n}(\gamma(s)) = Du(\gamma(s)) \cdot \mathbf{n}(s), \tag{10.6}$$

that is the *contrast of the image across the curve*. At a point $\gamma(s)$, $u_n(s)$ is larger if the magnitude of the gradient $|Du(\gamma(s))|$ is larger but also if the gradient is normal to the curve. The above Kass-Witkin-Terzopoulos, the Fua-Leclerc, and the Caselles-Kimmel-Sapiro contrast measures only take into account the magnitude of the gradient.

As a general remark on all variational snakes notice that since $g > 0$, the best snake energetic-wise is reduced to a single point at which the maximum magnitude of the image gradient is attained. Thus, in all snake models, local minima of the snake energy should be sought. The global ones are irrelevant. Such local minima usually depend on the initial position of the snake and the form of the contrast function g.

In their above-mentioned papers, Kimmel and Bruckstein made several important advances on the formulation of edge detectors and the snakes method. We can summarize the Kimmel-Bruckstein results as follows:

- Maximizers of $\int_0^{L(\gamma)} u_n(s)\,ds$ satisfy $\Delta u(\gamma(s)) = 0$, provided u_n does not change sign along the curve. This yields a variational interpretation of Hildreth-Marr edges linking them to the snakes method.
- Active contours can more generally be performed by maximizing a nonlinear function of contrast, $E(\gamma) = \int_0^{L(\gamma)} g(u_n(s))\,ds$, where g is even and increasing – a good example being $g(t) = |t|$. This is basically the energy (10.4) but where the isotropic contrast indicator $|Du(\gamma(s))|$ is replaced by the better term $u_n(s) = Du(\gamma(s)) \cdot \mathbf{n}(s)$ used in Montanari-Martelli method. The case $g(t) = t^2$ was actually considered earlier, in the founding Mumford-Shah paper [MS89]. They discovered that this functional has no minimizers because of the folding problem, which will be addressed later.
- Kimmel and Bruckstein considered maximizing the *average contrast*, namely

$$F(\gamma) = \frac{1}{L(\gamma)} \int_0^{L(\gamma)} g(u_n(s))\,ds, \tag{10.7}$$

where g is some increasing function. This is an improved version of the Fua-Leclerc functional.
- The evolution equation toward an optimum boundary can be written in much the same way as in the geodesic snake method.

All this looks good and well: The energy functional (10.7) is simple, it does not enforce *a priori* smoothness constraints, and it measures the real contrast. In the next section the main point left out, namely the shape of g, will be examined. *The form of g will be fixed by introducing a robustness requirement to contrast disparities along the contour.* Continuing, we will prove that this definition of snakes and the meaningful boundaries defined in Chapter 9 are essentially equivalent.

10.3 Choice of the Contrast Function g

Following Fua and Leclerc [FL90] and Kimmel and Bruckstein [KB01], we now consider the definition of a boundary as a closed curve along which the *average* contrast is maximal. Thus, we consider the energy given by (10.7):

$$F(\gamma) = \frac{1}{L(\gamma)} \int_0^{L(\gamma)} g(u_n)\,ds = \frac{E(\gamma)}{L(\gamma)},$$

where $L(\gamma)$ is the length of γ. The local maximization of (10.7) can be achieved by a numerical scheme detailed in Exercise 10.6.1. An example of an optimal snake with $g(t) = |t|$ is shown in Figure 10.1.

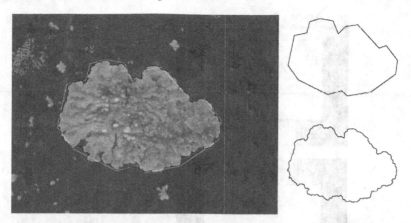

Fig. 10.1 An initial contour drawn by hand on the lichen image (curve top-right, in white on left image) and its final state (curve bottom-right, in black on the left image) computed with the average contrast snake model (10.7) for $g(t) = |t|$. The evolution allows an important improvement in the localization of the boundary of the object, as illustrated by the energy gain (360%, from 8.8 to 40.8). This shows the usefulness of the snake model as an interactive tool for contour segmentation.

We will show in this section that the shape of the contrast function g is extremely relevant. The energy $F(\gamma)$ to maximize is the average contrast $g(u_n)$ on the curve. $F(\gamma)$ increases when the curve is lengthened by adding a high-contrasted part or when it is shortened by removing a low-contrasted part. This remark raises a concern when the object boundary to be detected in the image has strong contrast variations. Indeed, if g has linear or superlinear growth, then the snake will tend to abandon the low-contrasted parts of the contour. Let us illustrate this fact by a little computation, associated to the crucial numerical experiment of Figure 10.2.

Consider a white square with side length 1 superimposed on a simple background image whose intensity is a linear function of x, varying from black to light gray. If a and b are the values of u_n on the left and right sides of the square respectively (we assume that $b < a$), and if γ is the boundary of the square, we have

$$F(\gamma) = \frac{E}{L}, \text{ with } L = 4 \text{ and } E = g(a) + g(b) + 2\int_0^1 g(a + (b-a)t)\,dt.$$

Now we would like to know if γ is an admissible final state of the snake model (i.e., a local maximum of F) or not. If we shrink the curve γ a little by "cutting" by $\varepsilon > 0$ the two right corners at 45 degrees, we obtain a curve γ_ε whose energy is

$$F_\varepsilon(\gamma) = \frac{E_\varepsilon}{L_\varepsilon}, \text{ with } L_\varepsilon = L - 4\varepsilon + 2\varepsilon\sqrt{2}$$

and $E_\varepsilon = E - 2\int_{1-\varepsilon}^1 g(a + (b-a)t)\,dt - 2\varepsilon g(b) = E - 4\varepsilon g(b) + o(\varepsilon).$

Since ε may be arbitrarily small, γ cannot be optimal if

$$\frac{L - L_\varepsilon}{L} > \frac{E - E_\varepsilon}{E}$$

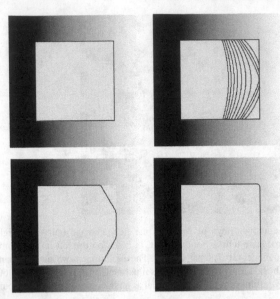

Fig. 10.2 Influence of the function g for a synthetic image. The snake model is applied to a synthetic image made of a bright square on a ramp background. Top-left: original contour. It can be detected as the unique optimal meaningful boundary of this image; see Chapter 9. Top-right: for $g(t) = |t|$, the snake collapses into a "flat curve" enclosing the left side of the square. Some intermediate states are shown to illustrate the snake evolution. Bottom-left: for $g(t) = |t|^{0.85}$, the snake converges to an intermediate state. Bottom-right: for $g(t) = |t|^{0.5}$, the snake hardly moves, which means that the initial contour is nearly optimal despite the large difference of contrast between the left side and the right side of the square. Contours with large variations of contrast are more likely to be optimal curves for low powers.

for $\varepsilon > 0$ small enough. Using the previous estimates of E_ε and L_ε and the fact that $E > 3g(b) + g(a)$, we can see that this condition is satisfied for ε small enough ($\varepsilon > 0$) as soon as

$$\frac{4 - 2\sqrt{2}}{4} > \frac{4g(b)}{3g(b) + g(a)},$$

which can be rewritten

$$\frac{g(a)}{g(b)} > \frac{2 + 3\sqrt{2}}{2 - \sqrt{2}} = 10.65\ldots$$

Hence, the ratio $g(a)/g(b)$ must be kept (at least) below that threshold and as small as possible in general in order to avoid the shrinkage of low-contrasted boundaries. If we choose a power function for g (i.e., $g(t) = |t|^\alpha$), this example is in favor of a small value of α, as illustrated in Figure 10.2.

More generally, the above argument shows that the contrast function g must increase as slowly as possible – in other words, be almost constant. On the other hand, all digital images are quantized and therefore have a minimal level of noise. This makes all gradient magnitudes below some threshold θ unreliable. We are led to the following requirement.

Flatness contrast requirement for snakes. *The contrast function for snake energy must satisfy, for some θ, the following:*

– *If $t \leq \theta$, $g'(t)$ is high.*
– *If $t \geq \theta$, $g(t)$ is flat and $g(t) \to g(\infty)$, with $g(\infty) < \infty$.*

Notice that the theory of Chapter 9 allows one to compute a meaningfulness contrast threshold θ as a function of the length of the curve. Let us focus on the special family of power functions $g(t) = |t|^{\alpha}$. Indeed, the power form yields a zoom invariant boundary detection method. When α is small enough these functions meet the above requirement and we can predict the following behaviors.

Experimental predictions. *Consider the average contrast energy*

$$F(\gamma) = \frac{1}{L(\gamma)} \int_0^{L(\gamma)} g(u_n(\gamma(s)))\,ds, \quad \text{with} \quad g(t) = |t|^{\alpha}. \tag{10.8}$$

– *When α is large, all snakes shrink from the parts of the boundaries with weaker gradient and replace them by straight parts.*
– *The smaller α is the better: Snakes are more stable and faithful to perceptual boundaries when $\alpha \to 0$.*

These predictions are confirmed by several numerical experiments. The first example is a synthetic image representing a white comb on a ramp background. In Figures 10.3 and 10.4 the evolution of the snake and its final state is represented for various values of α. As predicted, some teeth of the comb are not contrasted enough to be kept for $\alpha = 1$. A smaller value $\alpha = 0.4$ is required.

The same predictions are also confirmed when the snake model (10.8) is applied to a real photograph in Figure 10.5. The low-contrasted parts of the contour are kept only when the power α is small. They are replaced by straight lines when α is large.

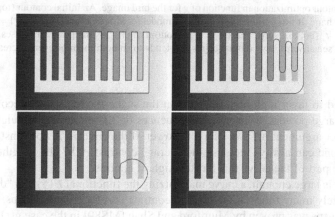

Fig. 10.3 Evolution of the snake for $g(t) = |t|$. The snake model is applied to a synthetic image of a bright comb on a ramp background. Top left: original contour (detected as the unique optimal meaningful boundary of this image). This contour loses progressively the comb's boundary on the low-contrast part. It stabilizes (bottom right) but misses three comb teeth.

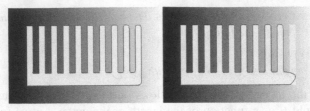

Fig. 10.4 Influence of the function g for a synthetic image. The snake model is applied to the same synthetic image of a bright comb on a ramp background. Left: $g(t) = |t|^{0.4}$. Right: $g(t) = |t|^{0.55}$. Compare with the result when $g(t) = |t|$ displayed on the bottom right of Figure 10.3. This experiment confirms the experimental prediction that a low power for g yields a better boundary detector.

Fig. 10.5 Contour optimization in function of g for the bird image. An initial contour (top, left) was first drawn by hand. It was optimized by the snake model for different functions g: $g(t) = |t|^{0.5}$ (top, right), $g(t) = |t|$ (bottom, left), and $g(t) = |t|^3$ (bottom, right). As the power increases, the snake becomes less sensitive to low contrast edges and tends to smooth them or even to create straight shortcuts.

The trend to favor high-contrast parts in the snake model, which becomes very strong for large powers, has some consequences on the numerical simulations and yield another argument in favor of a low power. If the contrast is not constant along the curve, one can always increase the average contrast (10.7) by lengthening the curve in the part of the curve that has the highest contrast.

When α is large enough, a curve maximizing the functional $F(\gamma)$ can "duplicate" itself infinitely many times in the highest-contrast part by creating cusps. This formation of cusps was proven by Mumford and Shah [MS89] in the case $g(t) = t^2$. We refer to Exercise 10.6.1 for details about the numerical scheme used for the experiments. In Figure 10.6 a "thick part" appears in the curve for $\alpha = 3$ and corresponds to such self-folding.

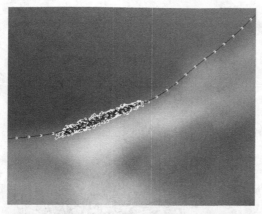

Fig. 10.6 Zoom on the curve duplication that appears for $g(t) = |t|^3$ in the highest-contrast part (rectangle drawn on the bottom-right image of Figure 10.5). The discretization of the snake (black curve) is shown by white dots.

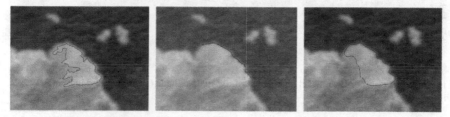

Fig. 10.7 Influence of the function g in the self-folding phenomenon. The initial boundary (left) is an optimal meaningful boundary of the lichen image. As in the square image (see Figure 10.2), the contrast along this curve has strong variations. The snake collapses into a self-folded "flat curve" with two cusps for $g(t) = |t|$ (middle) but remains a Jordan curve for $g(t) = \sqrt{|t|}$ (right).

The numerical experiment of Figure 10.7 is in some way the "real case" analog of Figure 10.2. In this experiment the initial region enclosed by the snake collapses into a zero-area region enclosed by a "flat curve".

10.4 Snakes Versus Meaningful Boundaries

In this section we would like to compare the snake model and the meaningful boundaries model (defined in Chapter 9, abbreviated MB). The optimal meaningful boundaries of the lichen image (resp. of the bird image) are shown in Figure 10.8 (resp. Figure 10.9).

The strength of snakes methods is that they combine gestalt principles of good continuation and of high contrast. Snakes models can find curves that are contrasted and interrupted, whereas the MB model cannot do the task of contour completion. The MB model will fail in cases where a point with a very low gradient occurs on the contour. However, although the MB model is fully automatic, an automatic

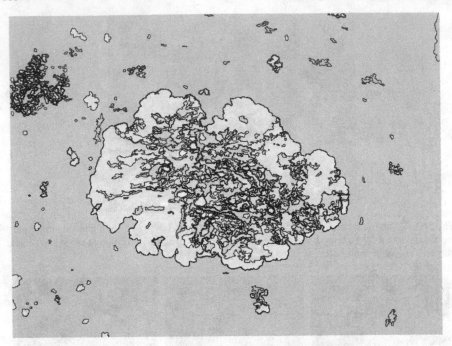

Fig. 10.8 The optimal meaningful boundaries of the lichen image (superimposed in light gray; see Figure 10.1).

Fig. 10.9 The optimal meaningful boundaries of the bird image (superimposed in light gray; see Figure 10.5).

parameterless algorithm is out of the scope of the snake method. An important user interaction is needed to ensure a good compromise between false detections and missed contours. The parameters to set are the initial contour and the parameters of the numerical scheme, including the gradient step and an initial image smoothing parameter (see Figure 10.10). Now, some recent methods do not have

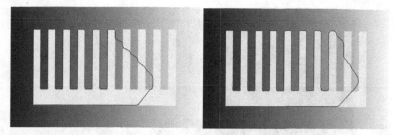

Fig. 10.10 Sensitivity of the snake model to the time step used for the gradient descent, see Exercise 10.6.1. The snake model ($g(t) = |t|$) is applied for several values of the gradient step δ: $\delta = 1$ (left), $\delta = 3$ (Figure 10.3, bottom-right), $\delta = 10$ (right). Due to the huge number of local maxima of the snake functional, the final result is very sensitive to the numerical implementation of the model, in particular to δ.

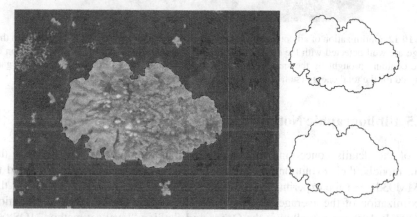

Fig. 10.11 Optimization of an optimal meaningful boundary by the snake model. The contour optimization brought by the snake model (here $g(t) = |t|$) is generally low when the contour is initialized as a contrasted level line (here an optimal meaningful boundary). In this experiment, the total energy is only increased by 17%, from 34.6 to 40.6.

these drawbacks anymore: An initialization is not needed and some other techniques of optimization such as graph cuts are used (see, for instance, [VC02], [BVZ01] or [XBA03])

In terms of *boundary optimization* (i.e., the refinement of a raw automatically detected or interactively selected contour), the snake model does not bring substantial improvements compared to the MB model. This fact was checked by applying the snake model to the contours detected by the MB model (Figures 10.11 and 10.12). The very few changes brought in these experiments by the snake evolution prove that the curves detected with the MB model are very close to local maxima of the snake model.

In conclusion, the snake model should be only used when interactive contour selection and optimization is required and when the sought object presents contrast inversions. In all other cases and in particular for automatic boundary detection, the meaningful boundaries method is more effective.

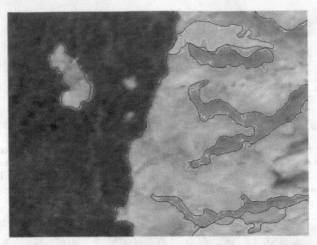

Fig. 10.12 Optimization of all optimal meaningful boundaries ($g(t) = \sqrt{|t|}$). The "objects" in this image are well detected with the optimal meaningful boundaries (in white). The optimization of these contours brought by the snake model (in black) is quite low, as shown by the little gain obtained for the total energy (sum of the energy of each curve): 7%, from 35.9 to 38.5.

10.5 Bibliographic Notes

All of the details concerning the formal computations needed for deriving the main models, their evolution equation, and steady-state equation can be found in [DMM03d]. For the experiments, we used a simple direct gradient descent for the maximization of the average contrast along the curve (energy given by Equation (10.7)). It does not actually use the Osher and Sethian "level set method" [OS88]. Indeed, the big advantage of these methods is to deal with topology changes of the snake during the minimization process. If the snakes can be replaced in most practical cases by level lines, the topological changes are simply handled by using their nested inclusion tree. More details about the "level set methodsn" in general with applications to imaging, vision, and graphics can be found in [OP03].

10.6 Exercise

10.6.1 Numerical Scheme

This exercise describes in detail a numerical scheme implementing the maximization of (10.7). For a non-Euclidean parameterization $\gamma(p) : [a,b] \to \mathbb{R}^2$, the energy we want to maximize writes

$$F(\gamma) = \frac{\int_a^b g\left(Du^\perp \cdot \frac{\gamma'(p)}{|\gamma'(p)|}\right) |\gamma'(p)| dp}{\int_a^b |\gamma'(p)| dp}. \tag{10.9}$$

1) Explain why it is better, rather than writing the Euler equation for (10.9) and *then* discretizing it, to discretize the energy and compute its exact derivative with respect to the discrete curve.

2) Let us suppose that the snake is represented by a polygonal curve $M_1, \ldots, M_n$ (either closed or with fixed endpoints). Show that a discretized version of the energy can be written $F = E/L$, where

$$L = \sum_i |\Delta_i| \qquad \text{with} \qquad \Delta_i = M_{i+1} - M_i$$

and

$$E = \sum_i g(t_i)|\Delta_i|,$$

with

$$t_i = w_i \cdot \frac{\Delta_i}{|\Delta_i|}, \quad w_i = Du^{\perp}(\Omega_i), \quad \text{and} \quad \Omega_i = \frac{M_i + M_{i+1}}{2}.$$

3) Show that the (exact) differentiation of F with respect to M_k is

$$\nabla_{M_k} F = \frac{1}{L}\left(\nabla_{M_k} E - F\nabla_{M_k} L\right),$$

with

$$\nabla_{M_k} L = \frac{\Delta_{k-1}}{|\Delta_{k-1}|} - \frac{\Delta_k}{|\Delta_k|},$$

$$\nabla_{M_k} E = v_k + v_{k-1} + z_{k-1} - z_k,$$

where

$$z_i = g'(t_i)w_i + [g(t_i) - t_i g'(t_i)]\frac{\Delta_i}{\|\Delta_i\|},$$

$$v_i = \frac{1}{2}g'(t_i)H(\Omega_i)\Delta_i$$

and the matrix H is defined by

$$H = \begin{pmatrix} -u_{xy} & u_{xx} \\ -u_{yy} & u_{xy} \end{pmatrix}.$$

4) Numerically, Du can be computed at integer points with a 3×3 finite differences scheme, and $D^2 u$ can be computed with the same scheme applied to the computed components of Du. This introduces a slight smoothing of the derivatives, which counterbalances slightly the strong locality of the snake model. These estimates at integer points are then extended to the whole plane using a bilinear interpolation.

To compute the evolution of the snake, one uses a two-step iterative scheme:

1. The first step consists in a reparameterization of the snake according to arc length. It can be justified in several ways: Aside from bringing stability to the scheme, it guarantees a *geometric* evolution of the curve, it ensures an homogeneous estimate of the energy, and it prevents singularities from appearing too easily. However, it does not prevent self-intersections of the curve.

2. The second step is simply a gradient evolution with a fixed step. If $(M_i^n)_i$ represents the (polygonal) snake at iteration n and $(\tilde{M}_i^n)_i$ its renormalized version after step 1, set

$$M_i^{n+1} = \tilde{M}_i^n + \delta \nabla_{\tilde{M}_i^n} F.$$

Implement this scheme.

Chapter 11
Clusters

11.1 Model

The perception of clusters or *proximity gestalt* is the first grouping process proposed by Wertheimer in his founding 1923 paper [Wer23]. Assume a set of n dots is drawn on a white sheet and those dots appear to be grouped in one or several clusters separated by blank spaces. Such clusters will be defined as very low-probability events in the a-contrario model that the dots have been uniformly and independently distributed over the white sheet. The area of the sheet is normalized to 1. Let σ be the area of a simply connected region A containing a k-dots cluster. The probability of observing at least k points among the n inside A is given by $\mathcal{B}(n, k, \sigma)$. Of course, A cannot be given *a priori* and the real event we are interested in is

There is a simply connected domain A, with area σ, containing at least k points.

The associated number of false alarms (NFA) is the expected number of such domains A, $N_{\mathcal{D}} \mathcal{B}(n, k, \sigma)$, where $\mathcal{D}$ is the set of all possible domains A and $N_{\mathcal{D}}$ is its cardinality.

11.1.1 Low-Resolution Curves

Testing too many domains would yield a large number of false alarms for all clusters and no detection. A set $\mathcal{D}$ of admissible domains as small as possible must be defined. Yet it must cope with all possible shapes for clusters. To that aim, we have to *sample* the set of simply connected domains by encoding at some precision their boundaries as Jordan curves. A possibility is to define these boundaries as discrete curves on a low-resolution grid of the image. The grid will be taken hexagonal because the number of polygons with given length is smaller than with a rectangular grid.

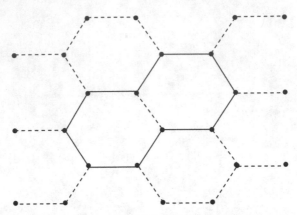

Fig. 11.1 The hexagonal mesh and an example of a closed low-resolution curve.

Definition 23. *Let us consider the hexagonal grid with step* $m \in \mathbb{R}$,

$$H_m = \left\{ m\left(k + \frac{l}{2}, l\frac{\sqrt{3}}{2} \right), \ (k,l) \in \mathbb{Z}^2, k - l \notin 3\mathbb{Z} \right\}.$$

A low-resolution curve of H_m *is a polygonal curve with l vertices* $M_1, M_2 \ldots, M_l$ *all in* H_m *and such that*

$$\forall i \in \{1, \ldots, l-1\}, \ |M_{i+1} - M_i| = m. \tag{11.1}$$

All of the vertices are different except $M_1 = M_l$.

An example of a closed low-resolution curve is given by Figure 11.1. A closed low-resolution curve is the boundary of a simply connected union of grid hexagons. Since we have no *a priori* idea of the size of the clusters to be detected, we need to consider several resolutions (mesh steps) $m \in M$, typically in geometric progression. Thus, N_D is the number of closed polygonal curves in all considered image grids.

Proposition 29 *Let m be a positive real number and l a positive integer. The number of closed low-resolution curves with l vertices lying on* $[0,1]^2 \cap H_m$ *is bounded from above by* $3 \cdot 2^{l-3} N_m$, *where the cardinality* N_m *of* $[0,1]^2 \cap H_m$ *satisfies*

$$N_m \simeq \frac{4}{3\sqrt{3}\,m^2}. \tag{11.2}$$

Proof — The number of closed low-resolution curves lying on $[0,1]^2 \cap H_m$, with l vertices and starting at a given point of $[0,1]^2 \cap H_m$, is bounded from above by $3 \cdot 2^{l-3}$ (3 possibilities for the second vertex, then at most 2 possibilities for each new vertex, the last one excepted). Hence, the number of closed low-resolution curves lying on $[0,1]^2 \cap H_m$ is bounded from above by $3 \cdot 2^{l-3} \cdot N_m$. Since the area of an hexagon of H_m is $\frac{3\sqrt{3}m^2}{2}$, the number of such hexagons in $[0,1]^2$ is asymptotically $\frac{2}{3\sqrt{3}m^2}$.

Each grid point being shared by three hexagons, the cardinality of $[0,1]^2 \cap H_m$ is asymptotically twice (6/3) the number of hexagons.   □

We will use the approximation given by (11.2) in the definition of false alarms. The allowed resolutions $m \in M$ will be taken in geometric progression with the ratio slighty above 1 (e.g., 1.2), with $\min(M)$ larger than the pixel size and $\max(M)$ lower than the image size, so that $|M|$ is actually a small number.

11.1.2 Meaningful Clusters

Definition 24. *We say that a group of k dots (among n) is an ε-meaningful cluster if there exists a closed low-resolution curve P of H_m ($m \in M$) with at most $l \leq L$ vertices, enclosing the k points in a domain with area σ, and such that*

$$\text{NFA}(l,m,k,\sigma) := \frac{L|M| \cdot 2^{l-1}}{m^2\sqrt{3}} \cdot \mathcal{B}(n,k,\sigma) \leq \varepsilon. \tag{11.3}$$

By the same standard arguments as for alignments or vanishing points and histogram modes, one obtains the following:

Proposition 30 *Consider n independent uniformly distributed points in the image. Then the expected number of ε-meaningful clusters is less than ε.*

11.1.3 Meaningful Isolated Clusters

It can also happen that a cluster is not overcrowded but only fairly isolated from the other dots. To take this event into account, let us introduce "thick" low-resolution curves obtained by dilating the low-resolution curves. The r-dilated D_rX of a plane set X is defined by

$$D_rX = \bigcup_{x \in X} B(x,r). \tag{11.4}$$

The events we now look for include the fact that no point should fall inside the dilated low-resolution curve defining the cluster domain A. In order to keep a finite count of all dilated low-resolution curves, one must constrain the dilation parameter r to belong to a fixed small set $R = \{r_1, \ldots, r_{|R|}\}$ in geometric progression.

Definition 25. *A group of k dots (among n) is called an ε-meaningful isolated cluster if there exists $r \in R$ and a closed low-resolution curve $P \in H_m$ ($m \in M$) with at most $l \leq L$ vertices, such that the r-dilated of P ($r \in R$) is an empty region with area σ', enclosing the k points in a domain with area σ and such that*

$$\text{NFA}(l,m,k,\sigma,\sigma') := \frac{L|M| \cdot |R| \cdot 2^{l-1}}{m^2\sqrt{3}} \cdot \sum_{i=k}^{n} \binom{n}{i} \sigma^i (1 - \sigma - \sigma')^{n-i} \leq \varepsilon. \tag{11.5}$$

Proposition 31 *For n independent and uniformly distributed points in the image, the expected number of ε-meaningful isolated clusters is less than ε.*

11.2 Finding the Clusters

11.2.1 Spanning Tree

According to Definition 24 (or 25), the detection of clusters could be realized by considering all possible (dilated or not) closed low-resolution curves and by counting the number of points falling inside. This is not feasible because the number of such curves is huge. The number of possible clusters is huge too. Thus, only subsets belonging to the spanning tree of S will be tested. We note respectively $\bar{B}(x,r)$ and $B(x,r)$ the closed and open disks with center x and radius r.

Definition 26. *Let $S \subset \mathbb{R}^2$ be a finite set of points. For any $A \subset S$, let*

$$\delta(A) = 2 \times \min\{r \geq 0 \mid \bigcup_{x \in A} \bar{B}(x,r) \text{ is connected}\},$$

with $\delta(A) = 0$ if $|A| \leq 1$. The spanning tree of S is defined as the unique tree $\mathrm{Span}(S)$ whose nodes are subsets of S and such that the following hold:

i) S is the root of $\mathrm{Span}(S)$.
ii) The children of a node N of $\mathrm{Span}(S)$ are (if any) the subsets of N that define the connected components of

$$\bigcup_{x \in N} B\left(x, \frac{\delta(N)}{2}\right).$$

From this definition, it is clear that the leaves of $\mathrm{Span}(S)$ are the elements of S and that $\delta(N)$ decreases as N goes from the root up to a leaf of the tree; so does N (with respect to inclusion), which proves that $\mathrm{Span}(S)$ has at most $2|S| - 1$ nodes.

11.2.2 Construction of a Curve Enclosing a Given Cluster

Proposition 32 *Let S be a finite subset of $\mathbb{R}^2$, $S' = S \cup {}^c[0,1]^2$ and N a node of $\mathrm{Span}(S)$. We note $\rho = \mathrm{dist}(N, S' \setminus N)$ and assume that $\rho > \delta(N)$. Then one connected component of the boundary of $D_{\rho/2}N$ is a Jordan curve $\mathcal{C} \subset [0,1]^2$ with finite length that encloses N and such that*

$$(D_r\mathcal{C}) \cap S' = \emptyset \tag{11.6}$$

for any $r \leq \rho/2$. No such curve enclosing N only can be found for $r > \rho/2$.

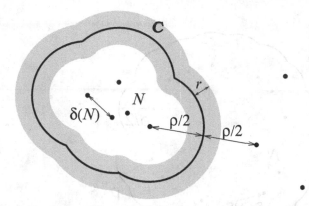

Fig. 11.2 A node N and its associated dilated enclosing curve (illustration of Proposition 32).

Proof — Let us consider the set

$$A = D_{\rho/2}N = \bigcup_{x \in N} B(x, \rho/2).$$

The boundary of this set is made of a finite disjoint union of Jordan curves (each having a finite length, because it is made of arcs of circles). Since $\rho > \delta(N)$, A is connected (but not necessarily simply connected); hence, one of these curves encloses all of the others (and, in particular, the whole set N). Let C be this curve (see Figure 11.2). Since $C \subset \partial A$, any point x of C satisfies $\mathrm{dist}(x, N) \geq \rho/2$, so that

$$(D_{\rho/2}C) \cap N = \emptyset. \tag{11.7}$$

Moreover, as

$$D_{\rho/2}C \subset D_{\rho/2}A = D_\rho N,$$

one has, by definition of ρ,

$$D_{\rho/2}C \cap (S' \setminus N) \subset (D_\rho N) \cap (S' \setminus N) = \emptyset.$$

With (11.7), this proves (11.6) for any $r \leq \rho/2$ (by monotonicity of the dilation operator). Notice that C may enclose a subset of S strictly larger than N (see Figure 11.3). Finally, let C be a Jordan curve enclosing N and N only. Since there exists a point $x \in N$ and a point $y \in S' \setminus N$ such that $\mathrm{dist}(x, y) = \rho$ (by definition of ρ), C has to intersect the segment $[x, y]$, so that for any $r > \rho/2$,

$$\{x, y\} \cap D_r C \neq \emptyset$$

and (11.6) cannot hold. $\qquad\square$

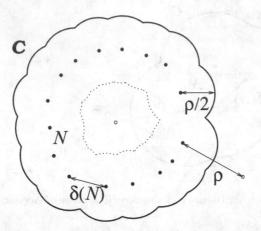

Fig. 11.3 C may enclose a subset of S strictly larger than N.

11.2.3 Maximal Clusters

As for the other gestalts, one can define maximal meaningful clusters by selecting local minima of the NFA with respect to inclusion (see Chapter 6).

Definition 27. *A node N of* $\mathrm{Span}(S)$ *is a maximal cluster if the following hold:*

– *For any child N' of N,* $\mathrm{NFA}(N') > \mathrm{NFA}(N)$.
– *For any parent N' of N,* $\mathrm{NFA}(N') \geq \mathrm{NFA}(N)$.

The same definition holds for isolated clusters.

Proposition 33 *Two distinct maximal clusters belonging to* $\mathrm{Span}(S)$ *cannot meet. The same property holds for isolated clusters.*

11.3 Algorithm

This section describes an algorithm achieving cluster detection according to the previous theory. We assume that $n = |S|$ is reasonably small, say $n \leq 1000$.

11.3.1 Computation of the Minimal Spanning Tree

The spanning tree associated to S can be computed with the following algorithm:

> initialization: each point of S is a tree
> while there remains more than one tree
>> find the 2 nearest trees and fuse them
> end while

When we fuse two trees A and B, they become the two children of a new node $A \cup B$ to which we attach a value δ, the distance between A and B (i.e., the minimum distance between a leaf of A and a leaf of B). The complexity of this step is $O(n^2 \log n)$ in the average, since the distances $(\text{dist}(s,t))_{(s,t) \in S^2}$ can be sorted once for all.

11.3.2 Detection of Meaningful Isolated Clusters

The next algorithm detects meaningful isolated clusters. A similar algorithm may be used for nonisolated clusters. Once the minimal spanning tree is computed, the cluster detection algorithm consists of a loop on all nodes of the tree inside which the NFA of each node is computed. For each node N:

> compute $A = D_{\rho/2}N$
> compute C, the enclosing boundary of A
> compute A', the domain enclosed by C
> if $(S \setminus N) \cap A' = \emptyset$:
>> for each resolution $m \in M$ such that $m \le \rho/4$:
>>> compute P, the closed low-resolution curve of H_m associated to C
>>> select r, the largest element of R such that $r + m \le \rho/2$
>>> compute $B = D_r P$
>>> compute $\sigma = \text{area}(A' \setminus B)$
>>> compute $\sigma' = \text{area}(B)$
>>> compute $\text{NFA}(|P|, m, |N|, \sigma, \sigma')$
>> end for
>> store the best NFA obtained for this node N
> end if

The above steps deserve some explanation:

- Since there is a minimal resolution (the smallest element of M), we consider the points of S as pixels of a discrete image with convenient resolution (in practice, the points of S are detected as points of a discrete image and the minimum resolution is taken slightly above the pixel size.) All erosions and dilations are then performed on this image and areas are computed by counting pixels on this image.
- The low-resolution curve $P \in H_m$ associated to the enclosing curve C is computed by projecting successive points of C on H_m (the Voronoï diagram associated to H_m is a triangular mesh), and then by processing the sequence obtained this way in order to obtain a low-resolution curve (no repetition of point, no cusp, no jump).

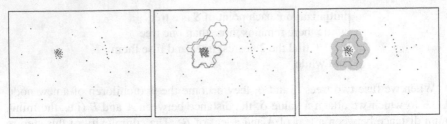

Fig. 11.4 Left: a set of dots. Middle: all 1-meaningful isolated clusters represented by their associated low-resolution curves (five clusters, $-\log_{10}(\text{NFA}) = 4.5/4.6/5.2/31/36$). Right: the only 1-meaningful maximal isolated cluster ($-\log_{10}(\text{NFA}) = 36$), represented by its associated low-resolution curve and empty region.

11.4 Experiments

In all of the following experiments, $L = 200$ and resolutions grow in geometric progression with ratio 1.2 from the pixel size (1) to the image size. R is an integer between 1 and 200.

11.4.1 Hand-Made Examples

The first cluster detection experiments are performed on two hand-drawn images (Figures 11.4 and 11.5).

11.4.2 Experiment on a Real Image

The black dots displayed in a satellite image are the centers of mass of contrasted small level sets. These level sets have at most 100 pixels and the gradient norm on their boundary is larger than 8. Figure 11.6 shows the meaningful and maximal meaningful clusters.

11.5 Bibliographic Notes

The detection of meaningful clusters is closely related to percolation theory (see, e.g., [Gri99], [SA94], [Mee96]). The aim of continuum percolation is, given a planar Poisson process with intensity λ of disks with radius r (possibly random), to study questions about the geometry of clusters such as: What is the probability distribution of the size of the finite connected components made by the disks? What is the threshold r_0 above which there is, with probability 1, an infinite connected component? See, for example, [Ale93], [QT97] or Hall's book [Hal88]. One can also replace

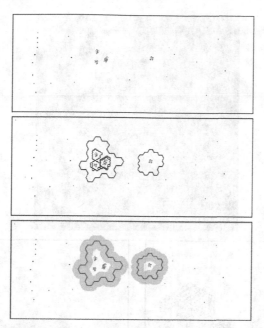

Fig. 11.5 Top: a set of 88 dots. Middle: all 1-meaningful isolated clusters represented by their associated low-resolution curve (six clusters, $-\log_{10}(\text{NFA}) = 0.20/0.43/2.1/17/29/43$). Bottom: the two maximal meaningful isolated clusters ($-\log_{10}(\text{NFA}) = 0.43$ (set of 12 dots) and $-\log_{10}(\text{NFA}) = 43$ (set of 45 dots)), represented by their associated low-resolution curve and empty region.

the disks by more general shapes (see, e.g., [RT02]). Most of the questions raised in continuum percolation remain open. The definition of an ε-meaningful group of dots given at the beginning of this chapter avoids all of the difficult questions of percolation by using low-resolution curves. Another topic in stochastic geometry directly related to cluster detection in dimension 2 is the theme of self-avoiding random walks. Estimates of the number of self-avoiding walks in any dimension are available. In dimension 2 for example, the number of self-avoiding walks with length l in a rectangular grid is asymptotically equivalent to k^l, where $2.5 \leq k \leq 2.7$. Numerical simulations indicate $k \simeq 2.64$. The number of self-avoiding polygons has the same order. These estimates given in [Law96] and [MS93] justify the choice of an hexagonal grid, since for such a grid, $k \leq 2$. We thank Pierre Calka, Gregory Randall, and Jérémie Jakubowicz for pointing out the above references to us.

The minimal spanning tree definition we used is not the usual one in geometric probability; see, for example, [PY03, Wu00, Ste02, Yuk00]. The tree used here does not give for each node the points that have connected its children, and thus no reconstruction is possible. This minimal spanning tree definition is closer to the definition of the connectedness tree used in statistics for hierarchical data classification [Sap90]. Unlike some problems in data analysis where the number of expected classes is known, here the number of clusters is *a priori* unknown and may even be zero.

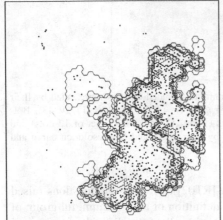 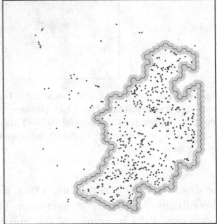

Fig. 11.6 Top: 531 dots detected in a 1000×1000 satellite image (superimposed). Middle: the 27 1-meaningful isolated clusters detected. Bottom: the only 1-meaningful maximal isolated cluster (491 dots, $-\log_{10}(\text{NFA}) = 199$).

The cluster detection theory presented here has been widely generalized and improved in [CDD⁺04]. In this huge paper, the authors extend the definition to any dimension. They improve the definition of maximal cluster, thus avoiding the obvious inaccuracy of the experiment in Figure 11.5 (in this experiment, a hierarchical group of made of three clusters is detected but the subgroups are not). The mentioned authors apply this clustering theory to shape recognition.

It often happens that the union of two clusters A and B is more meaningful that each one of them, but sometimes it makes sense to keep the pair of clusters separate. Thus, one needs to define the NFA of a pair of clusters, as compared to the NFA of the union of both clusters. Such a definition for pairs of clusters is proposed in [CDD⁺07]. A discussion and further improvement of this method can be found in [Pre06].

11.6 Exercise

11.6.1 Poisson Point Process

In this exercise, we will give formulas for the distance to the first, second, ..., k-th nearest neighbor in a Poisson point process. Such results and more properties about point processes can be found in the book by Stoyan and Stoyan [SS94] or in the book by Stoyan, Kendall, and Mecke [SKM87].

We first recall that an homogeneous planar Poisson point process of intensity $\lambda \in \mathbb{R}_+$ is characterized by the following properties:

- For a bounded Borel set B of $\mathbb{R}^2$, the number of points in B, denoted by $N(B)$, is a random variable with Poisson distribution of parameter $\lambda |B|$, where $|B|$ denotes the Lebesgue measure of B (i.e., its area). This means that

$$\forall k \in \mathbb{N}, \quad \mathbb{P}[N(B) = k] = e^{-\lambda |B|} \frac{(\lambda |B|)^k}{k!}.$$

- For all n integer and for all $B_1, B_2, \ldots, B_n$ bounded Borel sets such that $B_i \cap B_j = \emptyset$ for all $i \neq j$, the random variables $N(B_1), N(B_2), \ldots, N(B_n)$ are independent.

In what follows, we will consider an homogeneous planar Poisson point process of intensity $\lambda > 0$. We will denote by 0 the origin of $\mathbb{R}^2$.

1) Let $H_s(r)$ be the probability that the distance from 0 to its closest point of the point process is less than r. Prove that

$$H_s(r) = 1 - e^{-\lambda \pi r^2}.$$

(The function H_s is called the spherical contact distribution function.)

2) We now assume that the origin 0 is a point of the process. Let the probability that its first neighbor is a distance less than r be defined by

$$D_1(r) = \lim_{\varepsilon \to 0} \mathbb{P}[N(B(0, r) \setminus B(0, \varepsilon)) \geq 1 | N(B(0, \varepsilon)) = 1],$$

where $B(0, r)$ denotes the ball of radius r centered at the origin. Prove that

$$D_1(r) = 1 - e^{-\lambda \pi r^2}.$$

Remark: Notice that $D_1(r) = H_s(r)$. This is a particular property of the Poisson point process related to Palm distribution and Slivnyak's theorem (see [SS94]).

3) The distribution function of the first-neighbor distance is given by $D_1(r)$ and has density function $d_1(r) = D_1'(r)$. Prove that the mean and the variance of the first-neignbor distance are

$$m_1 = \frac{1}{2\sqrt{\lambda}} \quad \text{and} \quad \sigma_1^2 = \frac{1}{\pi \lambda} - \frac{1}{4\lambda}.$$

4) Prove that the distribution function of the k-th nearest-neighbor distance is

$$D_k(r) = 1 - \sum_{j=0}^{k-1} e^{-\lambda \pi r^2} \frac{(\lambda \pi r^2)^j}{j!}.$$

Compute the density function $d_k(r) = D'_k(r)$, and prove that the mean and the mode (defined as the argmax of the density function) of the k-th nearest-neighbor distance are

$$m_k = \frac{\Gamma(k + \frac{1}{2})}{\sqrt{\lambda} \pi (k-1)!} \quad \text{and} \quad r_k = \sqrt{\frac{k - \frac{1}{2}}{\pi \lambda}}.$$

Chapter 12
Binocular Grouping

12.1 Introduction

In this chapter, we present an application of meaningful events to binocular vision. Binocular vision, also called stereovision, is a major part of Computer Vision, as it theoretically allows us to reconstruct a 3-D representation of the world from two images taken from slightly different points of view. We focus here on the following problem: Given a set of n point matches between two images (i.e., a set of n points for each image with a one-to-one correspondence), can we decide when these points are the perspective projections on the two images of n physical points? Can we find the maximal group that satisfies this property? As we will see, the position of the point matches are constrained by the perspective projection as soon as $n \geq 8$ (for uncalibrated cameras). However, to prove in practice the existence of a rigid motion between two images, more than 8 point matches are desirable to compensate for the limited accuracy of the matches. In this chapter we describe a computational definition of rigidity and show how the Helmholtz principle can be applied to define a probabilistic criterion that rates the meaningfulness of a rigid set as a function of both the number of pairs of points (n) and the accuracy of the matches. This criterion yields an objective way to compare, say, precise matches of a few points and approximate matches of many points. It also yields absolute accuracy requirements for rigidity detection in the case of nonmatched points (i.e., when no one-to-one correspondence is available) and optimal values of n, depending on the expected accuracy of the matches and on the proportion of outliers. It can be used to build an optimized random sampling algorithm that is able to detect a rigid motion and estimate the fundamental matrix when the set of point matches contains up to 90% of outliers.

12.2 Epipolar Geometry

12.2.1 The Epipolar Constraint

Epipolar geometry elegantly describes the constraints satisfied by two perspective projections $m = (x\,y)^T$ and $m' = (x'\,y')^T$ of a physical 3-D point M on two different images. For pinhole cameras (with known or unknown internal parameters), the relation

$$(x'\,y'\,1)\,F\,(x\,y\,1)^T = 0 \tag{12.1}$$

holds, where F, the fundamental matrix, is a 3×3 matrix with rank 2 that depends on the rigid motion between the two image planes and on the camera parameters (position of the optical center, and relative pixel size along both axes). Equation (12.1) has the following interpretation: The point m' of the second image plane belongs to the epipolar line Fm whose equation $ax' + by' + c = 0$ is obtained from $m = (x, y)$ by

$$(a\,b\,c)^T = F\,(x\,y\,1)^T.$$

This epipolar line is nothing but the projection in the second image of the optical ray going from M to the optical center of the first camera. The position of m' on this line is related to the (unknown) depth of M, which is not involved in (12.1). The intersection of all epipolar lines is called the epipole: It is the projection in the second image of the optical center of the first camera.

The fundamental matrix can be factorized under the form $F = C'EC^{-1}$, where C and C' are 3×3 internal calibration matrices depending on the cameras only and E is the essential matrix defined from the motion parameters $T = (t_x\,t_y\,t_z)^T$ (3-D translation vector) and R (3-D rotation matrix) by

$$E = \begin{pmatrix} 0 & -t_z & t_y \\ t_z & 0 & -t_x \\ -t_y & t_x & 0 \end{pmatrix} R.$$

The essential matrix has, like F, a null singular value, but, in addition, its two other singular values are equal (see [FL01]). These notions are detailed in Exercise 12.8.1.

12.2.2 The Seven-Point Algorithm

The fundamental matrix has 7 degrees of freedom since it is a 9-dimensional data defined up to a multiplicative constant and satisfying a polynomial equation ($\det F = 0$). Hence, it can generally be estimated from 7 linear equations, using the so-called seven-point algorithm. From 7 point matches, (12.1) yields a 7×9 linear system $Af = 0$, where the fundamental matrix is written as a 9-dimensional vector f. If (F_1, F_2) is a basis of unit-norm solutions (in the nondegenerate cases, the kernel of A has dimension 2) with F_1 and F_2 written as 3×3 matrices, then the fundamental matrix is obtained as a solution $F = \alpha F_1 + (1 - \alpha)F_2$ of det $F = 0$. This condition requires, in order to determine α, the computation of the roots of a real cubic polynomial and may yield one or three solutions.

12.3 Measuring Rigidity

12.3.1 F-rigidity

The question we ask is the following: If $m_i = (x_i, y_i)$ and $m_i' = (x_i', y_i')$ are the n points found in the two images, how can we decide that each pair (m_i, m_i') represents the projection of a 3-D point in two different image planes? If this was exactly true, the epipolar constraint $(x_i' \, y_i' \, 1) F (x_i \, y_i \, 1)^T = 0$ should be satisfied for all i and for some 3×3 matrix F with rank 2, but in practice some location errors will occur and we have to measure how bad a correspondence is. We choose to measure the rigidity of a matching set $S = \{(m_i, m_i')\}_{i=1..n} \in (I \times I')^n$ (I and I' represent the (convex) domains of the two images) with the following physical criterion.

Definition 28. *Let F be a 3×3 matrix with rank 2. The F-rigidity of a set S of point matches between two images is*

$$\alpha_F(S) = \frac{2D'}{A'} \max_{(m,m') \in S} \text{dist}(m', Fm), \tag{12.2}$$

where Fm is the epipolar line associated by F to m, dist is the Euclidean distance, A' is the area of the second image domain, and D' is its diameter.

We will justify later the normalization coefficient $2D'/A'$, but we notice that it yields a scale-independent definition of rigidity. Contrary to classical criteria, we use the l^∞ norm (maximum) instead of the l^2 norm (sum of squares). This has the advantage of being extremely selective against outliers and simplifies most of the computations that will follow. However, the same approach could as well be applied to a more usual l^2 criterion by using the Central Limit Theorem. As usual, (12.2) may be symmetrized by using

$$\tilde{\alpha}_F(S) = \max_{(m,m') \in S} \max \left(\frac{2D'}{A'} \text{dist}(m', Fm), \frac{2D}{A} \text{dist}(m, F^T m') \right) \tag{12.3}$$

(A is the area of the first image domain and D is its diameter).

We could define a meaningful point correspondence as a matching set S for which $\alpha_F(S)$ is too small (for some F) to be reasonably explained by randomness. More precisely, if we suppose that the points of S are randomly, uniformly, and independently distributed in $(I \times I')^n$, then we can define the probability $q(t) = P[\inf_F \alpha_F(S) \leq t]$. A meaningful point correspondence could then be defined as a matching set S for which $q(\inf_F \alpha_F)$ is very small, say less than 10^{-3}. However, the computation of the *probability q* of the meaningful event is as usual very difficult. This is why we will use a NFA model based on *expectation*. Instead of controlling the probability of a false detection (a false detection is a rigid set detected in a random distribution of points), we measure the expected number of false detections, which leads to more tractable computations (an explicit illustration of this is given in Exercise 3.4.1 on birthday dates). As we will see now, this is possible as soon as we quantize the number of tests (each test tries a candidate for F).

12.3.2 A Computational Definition of Rigidity

According to the previous remark, the rigidity of a matching set S should be defined by $\inf_F \alpha_F(S)$ –that is, by minimizing the matching error among all possible F. However, this definition is too difficult to manage. A more computational point of view must be adopted.

Definition 29. *A matching set* $S = \{(m_i, m'_i)\}_{i=1..n}$ *is* α-*rigid if there exists a fundamental matrix* F *associated to a subset of 7 matchings of* S *such that* $\alpha_F(S) \leq \alpha$.

Note that we could have used $\tilde{\alpha}_F(S)$ as well in this definition. From this definition, the *rigidity* of S can be defined as the least α for which S is α-rigid. Finding this exact minimum requires the computation of up to $3\binom{n}{7}$ fundamental matrices since, as we saw previously, the seven-point algorithm may produce up to three solutions. Such a computation becomes rapidly unfeasible when n grows larger than 25 (see Table 12.1). However, since the minimum value of α is unlikely to be isolated, for large n it is enough in practice to compute the minimum of $\alpha_F(S)$ over a small proportion of possible subsets S, which justifies Definition 29.

In comparison with the ideal approach that would consist in estimating the best F among all possibilities (and not only among solutions of the seven-point algorithm),

Table 12.1 Number of transforms required to define the exact rigidity of a set of n point matches and the associated computation time on a 1-GHz PC laptop, estimated on a basis of 30,000 fundamental matrices computed per second (see Section 12.5). The rapidly increasing computational cost justifies the less systematic Definition 29 of α-rigidity

n	Number of transforms, $3\binom{n}{7}$	Expected computation time
8	24	0.0008 s
10	360	0.012 s
15	19,305	0.64 s
20	232,560	7.7 s
25	1,442,100	48 s
30	6,107,400	3 min 30 s
50	$3\ 10^8$	2 h 45 min
100	$4.8\ 10^{10}$	18 days

we have reduced the number of possible F in the spirit of the RANSAC paradigm (see Section 12.7). This is not too restrictive though, since if there exists a good correspondence between the point matches of S, it is likely that the seven-point algorithm will produce a good approximation of the "ideal" F (i.e., minimizing $\alpha_F(S)$) for some subset of seven point matches of S.

12.4 Meaningful Rigid Sets

12.4.1 The Ideal Case (Checking Rigidity)

Now that we have given a definition of approximate rigidity, we use the notion of an ε-meaningful event to reduce the set parameters (n, the size of S, and α, the rigidity threshold) to a single intuitive parameter, the expected number of false alarms. To this aim let us take the usual *a contrario* Helmholtz assumption that the points m_i and m'_i are uniformly and independently distributed on I and I'.

Proposition 34 *A matching set S of size n ($n \geq 8$) is ε-meaningful as soon as it is α-rigid with*

$$\varepsilon_1(\alpha, n) := 3 \cdot \binom{n}{7} \cdot \alpha^{n-7} \leq \varepsilon. \tag{12.4}$$

The number ε_1 measures the meaningfulness (from the rigidity viewpoint) of a given set of point matches. Hence, Proposition 34 not only provides an intuitive threshold for rigidity detection but also gives a way to compare the meaningfulness of two set. Since ε_1 measures the expected number of α-rigid sets of size n for random points, the smaller ε_1 is, the better the accuracy of S is. We can also notice that Proposition 34 remains true when α-rigidity is defined with $\tilde{\alpha}_F$ (instead of α_F) since $\tilde{\alpha}_F \leq \alpha_F$. We keep using α_F in the following because we did not see how to significantly improve the meaningfulness thresholds in the symmetric case (in particular, because some Fs are such that $\tilde{\alpha}_F = \alpha_F$).

Proof of Proposition 34 — Let $T \subset S$ be a set of 7 point matches, and let F be one of the associated rank 2 fundamental matrix. If (m_i, m'_i) belongs to T, then the distance from m'_i to the epipolar line $l = F m_i$ is zero by construction. If it belongs to $S - T$, then m'_i and l are independent, and since the intersection of l with I' has length less than D, the t-dilated of l in I' (i.e., the set of points of I' whose distance from l is lower than t) has area less than $2tD$ and the probability that $\text{dist}(m'_i, l) \leq t$ is less than $2tD/A$. Consequently, the probability that $\alpha_F(S) \leq \alpha$ is less than α^{n-7}. Now, the number of subsets T of S with 7 elements is $\binom{n}{7}$, and since each of these subsets can produce at most 3 matrices F, the maximum number of fundamental matrices we can build from S with the seven-point algorithm is $3 \cdot \binom{n}{7}$. Hence, for every α satisfying (12.4), any α-rigid set of n point matches is ε-meaningful. $\square$

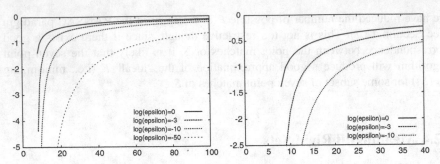

Fig. 12.1 Detectability curves in the $(n, \text{Log}\,\alpha)$ plane according to Equation (12.4) for several meaningfulness values ε_1. A rigidity value of $\text{Log}\,\alpha = -2$ corresponds to a maximum epipolar distance equal to 3.5 pixels on a 1000×1000 image (all logarithms taken in base 10).

Equation (12.4) is very encouraging for the detection of rigidity in the case of already matched points. Indeed, as n grows, the left term goes quickly to zero and a set of point matches is not required to be α-rigid with α very small to be detected as rigid, provided that it contains enough points. The link between α, n, and ε is illustrated on Figure 12.1. We can see, for example, that even with a low accuracy (3.5 pixels on a 1000×1000 image), 10 points are enough to define a meaningful rigid set ($\varepsilon = 10^{-3}$).

More precisely, when n tends to infinity, (12.4) writes

$$\text{Log}\,\alpha \leq -7\frac{\text{Log}\,n}{n} + \frac{C + \text{Log}\,\varepsilon}{n} + \underset{n \to +\infty}{O}\left(\frac{\text{Log}\,n}{n^2}\right), \qquad (12.5)$$

where $C = \text{Log}(7!/3) = 3.22...$ (here and in all of the following, Log means the logarithm in base 10).

Proof of (12.5) — We have, when $n \to +\infty$,

$$\text{Log}\binom{n}{7} = 7\text{Log}\,n - \text{Log}(7!) + O\left(\frac{1}{n}\right),$$

so that if $3\binom{n}{7}\alpha^{n-7} = \varepsilon$, then

$$\text{Log}\,\alpha = \frac{1}{n-7}\left(\text{Log}\,\varepsilon + \text{Log}\,\frac{7!}{3} - 7\text{Log}\,n + O\left(\frac{1}{n}\right)\right)$$

$$= -7\frac{\text{Log}\,n}{n} + \frac{C + \text{Log}\,\varepsilon}{n} + O\left(\frac{\text{Log}\,n}{n^2}\right) \qquad (12.6)$$

with $C = \text{Log}(7!/3)$. $\square$

12.4.2 The Case of Outliers

Up to now, we have studied the meaningfulness of the whole set of point matches. This is not very realistic since, in practice, the presence of outliers (badly located points or false matches) cannot be avoided. Moreover, Definition 29 is very sensitive to outliers, since it measures rigidity by a *maximum* of epipolar distances. We now consider rigidity detection for a subset S' of the set S of all point matches. Compared to the previous case, we now have to count, in the expected number of false alarms, the number of possible choices for k (i.e., $n - 7$) and the number $\binom{n}{k}$ of possible subsets of S with size k.

Proposition 35 *A set $S' \subset S$ of k point matches among n is ε-meaningful as soon as it is α-rigid with*

$$\varepsilon_2(\alpha, n, k) := 3(n - 7) \cdot \binom{n}{k} \cdot \binom{k}{7} \cdot \alpha^{k-7} \leq \varepsilon. \tag{12.7}$$

Figure 12.2 illustrates the relation between n, $p = 1 - k/n$ and α for $\varepsilon = 10^{-3}$. In Figure 12.3, the meaningfulness achieved for fixed values of n and α is presented as a function of p.

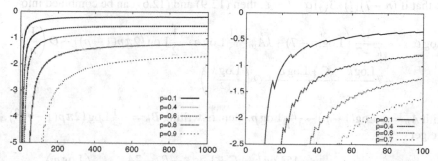

Fig. 12.2 Detectability curves ($\varepsilon_2 = 10^{-3}$) in the $(n, \mathrm{Log}\,\alpha)$ plane for different proportions p of outliers (see Equation (12.7)). The curves present small oscillations because $p = 1 - k/n$ cannot be exactly achieved by reason of the discrete nature of k and n (this could be avoided by interpolating the binomial coefficients with the Gamma function).

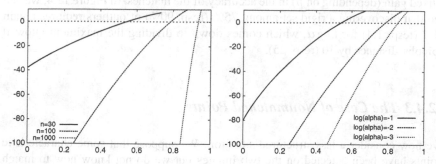

Fig. 12.3 Log-meaningfulness $\mathrm{Log}(\varepsilon_2)$ as a function of the proportion p of outliers (Equation (12.8)). Left: $\alpha = 10^{-2}$ and n varies. Right: $n = 100$ and α varies.

As before, we can derive asymptotic estimates from (12.7). If $p = 1 - k/n$ is the (fixed) proportion of outliers, then assuming $0 < p < 1$ and $n \to \infty$, we can rewrite (12.7) as

$$\text{Log}\,\alpha \leq -A_p - \frac{15\,\text{Log}\,k}{2k} + \frac{C + \text{Log}\,\varepsilon + B_p - 7A_p}{k} + \underset{n\to+\infty}{O}\left(\frac{\text{Log}\,n}{n^2}\right), \quad (12.8)$$

with $A_p = -\text{Log}(1-p) - \frac{p}{1-p}\,\text{Log}\,p$ and $B_p = -\frac{1}{2}\text{Log}\left(2\pi p(1-p)^2\right)$.

Proof of (12.8) — If $k = (1-p)n$ (p fixed, $0 < p < 1$) and $n \to +\infty$, Stirling's formula yields

$$\text{Log}\binom{n}{k} = -n\big(p\,\text{Log}\,p + (1-p)\,\text{Log}(1-p)\big)$$
$$- \frac{1}{2}\text{Log}\,n - \frac{1}{2}\text{Log}\left(2\pi p(1-p)\right) + O\left(\frac{1}{n}\right), \quad (12.9)$$

so that if $(n-7)\binom{n}{k} \cdot 3\binom{k}{7}\alpha^{k-7} = \varepsilon$, then (12.9) and (12.6) can be combined into

$$\text{Log}\,\alpha = -\frac{1}{k-7}\left(\text{Log}(n-7) + kA_p - \frac{1}{2}\text{Log}\,n - \frac{1}{2}\text{Log}\left(2\pi p(1-p)\right) + O\left(\frac{1}{n}\right)\right)$$
$$-7\frac{\text{Log}\,k}{k} + \frac{C + \text{Log}\,\varepsilon}{k} + O\left(\frac{\text{Log}\,k}{k^2}\right),$$

with $A_p = -\text{Log}(1-p) - \frac{p}{1-p}\,\text{Log}\,p$. Hence, writing $B_p = -\frac{1}{2}\text{Log}\left(2\pi p(1-p)^2\right)$, we get

$$\text{Log}\,\alpha = -A_p - \frac{15\,\text{Log}\,k}{2k} + \frac{C + \text{Log}\,\varepsilon + B_p - 7A_p}{k} + O\left(\frac{\text{Log}\,n}{n^2}\right).$$

$\square$

Compared to (12.5), the main difference lies in the leading term $-A_p$ that forces a fixed gain (depending on p) in the accuracy of the matches. In Figure 12.4, we can see that discovering a rigid set among 75% (resp. 90%) of outliers requires a gain of 1 (resp. 1.4) for $\text{Log}\,\alpha$, which comes down to dividing the maximum allowed epipolar distance by 10 (resp. 25).

12.4.3 The Case of Nonmatched Points

We now consider a less structured situation: We suppose that some characteristic points have been detected on the two images but we do not know how to match them together. In this case, we have to take into account the fact that *two* subsets of k points have to be chosen and that $k!$ possible permutations remain possible to define point matches between these two subsets.

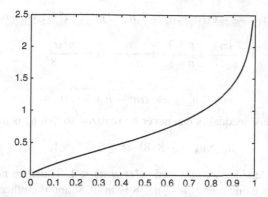

Fig. 12.4 The required rigidity gain A_p in function of the proportion of outliers (p).

Proposition 36 *Let $P = \{m_i\}_{i=1..n}$ and $P' = \{m'_i\}_{i=1..n}$ be the points detected on each image. A matching set S of size k formed by disjoint pairs taken from $P \times P'$ is ε-meaningful as soon as it is α-rigid with*

$$\varepsilon_3(\alpha, n, k) := 3(n-7) \cdot \binom{n}{k}^2 \cdot k! \cdot \binom{k}{7} \cdot \alpha^{k-7} \le \varepsilon. \qquad (12.10)$$

We have assumed for simplicity that $|P| = |P'|$, but in the general case where $|P| = n$ and $|P'| = n'$, we would obtain the condition

$$3 \min(n-7, n'-7) \cdot \binom{n}{k} \cdot \binom{n'}{k} \cdot k! \cdot \binom{k}{7} \cdot \alpha^{k-7} \le \varepsilon.$$

The condition (12.10) is very different from the previous case (12.7), because the matching pairs now have to be found. In particular, even for $n = k$ (all points matched), the function $n \mapsto \varepsilon_3(\alpha, n, n)$ is *increasing* when n grows to infinity. This means that for a fixed rigidity α, no meaningful rigid set can be found beyond a certain value of n. The following proposition states this more precisely.

Proposition 37 *If $\varepsilon_3(\alpha, n, n) < 1$, then*

$$\alpha < \frac{1}{32} \qquad \text{and} \qquad n < \frac{e}{\alpha} + \underset{\alpha \to 0}{O}(\mathrm{Log}\, \alpha),$$

and the minimum of $n \mapsto \varepsilon_3(\alpha, n, n)$ is obtained for a unique $n = \bar{n}$ such that

$$\bar{n} = \frac{1}{\alpha} - 8 + \underset{\alpha \to 0}{O}(\alpha).$$

Moreover, one has

$$\mathrm{Log}\, \varepsilon_3(\alpha, \bar{n}, \bar{n}) = -\frac{\mathrm{Log}\, e}{\alpha} - \frac{3}{2} \mathrm{Log}\, \alpha - C' + \underset{\alpha \to 0}{O}(\alpha), \qquad (12.11)$$

with $C' = \mathrm{Log}\, \frac{7!}{3} - \frac{1}{2} \mathrm{Log}(2\pi) = 2.82...$ and $\mathrm{Log}\, e = 0.434...$

Proof — Let us write $u_n = \varepsilon_3(\alpha, n, n) = 3(n-7)n!\binom{n}{7}\alpha^{n-7}$. We have

$$\frac{u_n}{u_{n-1}} = \frac{n-7}{n-8} \cdot n \cdot \frac{n}{n-7} \cdot \alpha = \frac{n^2\alpha}{n-8},$$

so that

$$u_n \leq u_{n-1} \Leftrightarrow \alpha n^2 - n + 8 \leq 0.$$

If $\alpha \geq 1/32$, this last inequality can never be realized so that u_n is increasing and

$$u_n \geq u_8 = 3 \cdot 8 \cdot 8! \cdot \alpha \geq 6 \cdot 7! \gg 1,$$

which is impossible since we assumed $\varepsilon_3(\alpha, n, n) < 1$. Thus, we necessarily have $\alpha < 1/32$, and since one root of $n^2 - n + 8$ is in $]0, 8[$ and the other $(\bar{n})$ larger than 8, u_n is decreasing for $n \leq \bar{n}$ and increasing for $n \geq \bar{n}$, with

$$\bar{n} = \frac{1 + \sqrt{1 - 32\alpha}}{2\alpha} = \frac{1}{\alpha} - 8 + \underset{\alpha \to 0}{O}(\alpha).$$

Moreover, when $n \to \infty$ one has

$$\mathrm{Log}\, u_n = n \,\mathrm{Log}\, \frac{\alpha n}{e} + O(\mathrm{Log}\, n),$$

so that the condition $u_n \to 1$ imposes

$$\frac{\alpha n}{e} = 1 + O\left(\frac{\mathrm{Log}\, n}{n}\right),$$

which means that

$$n = \frac{e}{\alpha} + \underset{\alpha \to 0}{O}(\mathrm{Log}\, \alpha).$$

Finally, we have

$$\mathrm{Log}\, u_n = -\mathrm{Log}\, \frac{7!}{3} + \mathrm{Log}\, n + n \,\mathrm{Log}\, \frac{n}{e} + \frac{1}{2} \mathrm{Log}(2\pi n)$$

$$+ 7\,\mathrm{Log}\, n + (n-7)\,\mathrm{Log}\, \alpha + \underset{n \to +\infty}{O}\left(\frac{1}{n}\right),$$

and since $\alpha = \frac{1}{\bar{n}}(1 - \frac{8}{\bar{n}})$,

$$\mathrm{Log}\, u_{\bar{n}} = \frac{3}{2} \mathrm{Log}\, \bar{n} - (\bar{n} + 8)\,\mathrm{Log}\, e - \mathrm{Log}\, \frac{7!}{3} + \frac{1}{2} \mathrm{Log}\, 2\pi + O\left(\frac{1}{\bar{n}}\right)$$

$$= -\frac{\mathrm{Log}\, e}{\alpha} - \frac{3}{2} \mathrm{Log}\, \alpha - C' + \underset{\alpha \to 0}{O}(\alpha)$$

with $C' = \mathrm{Log}\, \frac{7!}{3} - \frac{1}{2} \mathrm{Log}\, 2\pi$. $\square$

This proposition calls for several comments. First, it yields a universal threshold $(1/32)$ on the rigidity coefficient α, above which a set cannot be meaningfully rigid. Second, for any expected rigidity α, it suggests an optimal number $n = \bar{n}(\alpha)$ of characteristic points to be found on each image and guarantees that rigidity detection

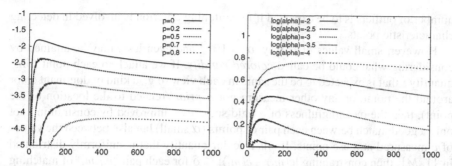

Fig. 12.5 Detectability curves ($\varepsilon_3 = 10^{-3}$) in the $(n, \text{Log}\,\alpha)$ plane for several values of p (left) and in the (n, p) plane for several values of α (right).

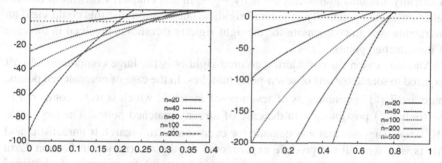

Fig. 12.6 Log-meaningfulness $\text{Log}(\varepsilon_3)$ as a function of the proportion p of outliers, for several values of n. Left: $\alpha = 10^{-2.5}$; right: $\alpha = 10^{-4}$.

is possible for that n since the right-hand term of (12.11) is negative for α small enough.

The situation is similar in the case of outliers. In Figure 12.5 (right), we can see that given a precision α and a meaningfulness threshold (e.g., $\varepsilon_3 = 10^{-3}$), the maximum allowed amount p of outliers is not an increasing function of n. It increases up to an optimal value of n and then decreases for larger values of n. For p fixed, the required rigidity threshold α behaves in a similar way, and the less restrictive threshold is obtained for a particular value of n (Figure 12.5, left).

For a practical[1] value of $\alpha = 10^{-2.5}$, we can see on Figure 12.6 (left) or on Figure 12.5 (right) that the maximum allowed amount of outliers is around $p = 0.25$ and is attained around $n = 80$. Large values of p can be attained for $\alpha = 10^{-4}$ for example (see Figure 12.6, right), but this would require an unrealistic precision (0.035 pixel in a 1000×1000 image) in the localization of the characteristic points. This outlines an absolute limit of rigidity detection in the case of nonmatched points and suggests that in most situations, rigidity detection is not possible, since a large

---

[1] $\alpha = 10^{-2.5}$ means a precision of about 1 pixel on a 1000×1000 image (see Equation (12.2)). Achieving smaller values is not impossible but requires an excellent compensation of the camera distortions and an accurate subpixel registration algorithm.

number of outliers is to be expected if no matching criterion is involved in detecting characteristic points.

However, small values of α (say, $\alpha = 10^{-4}$ and even less) may be attained by considering what could be called *colored rigidity*. If we attach to each point some quantity c that is expected to be the same in each view (e.g., a kind of dominant color around the point, or any other measure not *a priori* related to the location of the point), then the meaningfulness of a rigid set can be improved by constraining not only a good match between each pair of points (α small) but also between the values of c measured at these points. If c is a scalar quantity that is uniformly distributed in $\{1..M\}$, then constraining $|c(m_i) - c(m_i')| \leq \delta$ for each pair (m_i, m_i') of matching points allows us to multiply the meaningfulness coefficient ε_3 by $\left(\frac{2\delta+1}{M}\right)^k$, which practically amounts to multiplying α by $\frac{2\delta+1}{M}$. In this chapter, we will not investigate further this possibility of colored rigidity, but mention that this could yield an interesting way to compensate for the tight rigidity thresholds involved in the case of nonmatched points.

Another reason for considering colored rigidity is the huge computational cost required to search for sets of seven-point matches. In the case of nonmatched points, there are $7!\binom{n}{7}^2$ possible sets of seven-point matches, which is to be compared to the (only) $\binom{n}{7}$ possible sets in the case of already matched points. The large ratio $7!\binom{n}{7} \simeq n^7$ suggests that a systematic (or even stochastic) search is unrealistic, and this is why we will not produce experiments for the ε_3 criterion in Section 12.6. However, we do think that the ε_3 criterion may be useful, for example, if additional information (colored rigidity mentionned above, or local image comparison near the points to match) constrains the search for matches to a limited number of cases.

12.4.4 A Few Remarks

Before we describe the algorithms and present experiments, let us discuss the validity of the three models presented. The first model (ε_1) concerns only the case when all point matches are to be kept in the rigidity detection. Even if it is known that no outliers are present, the use of this model is questionable (i.e., the second model should be preferred) since the less precise point matches may reduce the meaningfulness of the whole set and compromise the precision of F. Hence, this first model is mainly of pedagogic value.

The second model (ε_2) is applicable when point matches are available between the two images, provided that no location criterion was involved in the matching process. If, for example, each point m_i has been matched to the nearest m_j' in the second image (for some distance that may not be the usual Euclidean distance), which supposes that some *a priori* assumption is made on the relative camera positions, then a meaningful rigid set may be found by accident with the ε_2 criterion, since the uniform model against which rigid sets are detected is no longer valid. In this case, or when point matches are not known, the third model (ε_3) should be used.

The meaningfulness thresholds we derived should be distinguished from the algorithms used for rigidity detection. Even if the meaningfulness thresholds count *all* possible transforms F obtained from 7 point matches, which may represent a huge number (especially for ε_3), a selective exploration of these transforms can be used to detect rigidity. If, for example, some restrictions are known about the epipolar geometry (this may be the case for a stereo pair, e.g.), then this knowledge might be very useful to speed up rigidity detection. All kinds of optimized strategy for rigidity detection may be used, provided that all *possible* transforms are counted in the derivation of the meaningfulness thresholds, as we did for ε_3.

We also would like to point out that even if our model uses a uniform distribution of points to derive the meaningfulness coefficient, this is not an assumption to be realized by the observed data since we detect rigid sets *against* this uniform model. If, for example, all points of the second image are concentrated in a very small region, then this particular concentration (which is in some way meaningful) might raise the detection of a rigid set. If a symmetrical definition of rigidity is used as suggested in Section 12.3.2, then this might only happen if this concentration also occurs in the first image. In this case, the existence of a rigid set still makes sense, although it would probably not be the *simplest explanation* of the observed distribution of points. This suggests that a similar approach should also be applied for the detection of simpler motions (given by homographies, affine or even Euclidean transforms), in the spirit of Occam's razor principle (*the simplest explanation is the best*). The simplest explanation could then be found by using a Minimum Description Length [Ris83, Ris89] criterion. Another possibility would be to detect rigidity against a more general model of point distribution allowing a certain amout of clustering, but it is not clear that the new obtained thresholds would be significantly different.

12.5 Algorithms

12.5.1 Combinatorial Search

The seven-point algorithm was implemented in C using the Singular Value Decomposition algorithm from Numerical Recipes [PTVF88] and a classical explicit solver for the third-degree equation raised by the rank 2 constraint. The resulting algorithm is able to test 10,000 seven-point matches per second (i.e., between 10,000 and 30,000 fundamental matrices) on a 1-GHz PC laptop.

This algorithm proceeds as follows. When considering a fundamental matrix F given by a subset T of 7 point matches of S, compute for each of the $n - 7$ remaining pairs the normalized distance α_i between m_i' and the epipolar line Fm_i. Then sort these distances in the increasing order and obtain for each value of k between 8 and n the subset $S_k(F)$ of size k such that $T \subset S_k \subset S$ and $\alpha_F(S_k)$ is minimal. Since $\varepsilon_2(\alpha, n, k)$ is increasing with respect to α (see Equation (12.7)), $S_k(F)$ is also the most meaningful (with respect to F) rigid set of size k containing T. Then find, among the sets $\left(S_k(F)\right)_{k=8..n}$, the most meaningful one (i.e., the one minimizing ε_2), written $\bar{S}(F)$.

The deterministic algorithm works as follows: For each subset T of 7 point matches of S and each of the (possibly 3) fundamental matrices F associated to T, compute $\bar{S}(F)$. The result of the algorithm is simply the most meaningful set $\bar{S}$ encountered in this systematic search. Also compute, for each value of k, the most meaningful set *with size k*, that is the most meaningful set S_k among the $S_k(F)$. By construction, this function of k is minimal when $S_k = \bar{S}$.

12.5.2 Random Sampling Algorithm

The above algorithm must be abandoned when $n \geq 25$ in favor of a stochastic version, as shown by Table 12.1. Following the random sampling consensus, similar computations of rigidity and meaningfulness can be run only for certain fundamental matrices F obtained from random sets of seven-point matches. The algorithm can be written as follows:

> set $\bar{\varepsilon} = +\infty$
> repeat
> > generate a random set T of 7 point matches
> > for each fundamental matrix F associated to T
> > > compute the most meaningful rigid set $\bar{S} = \bar{S}(F)$ associated to F
> > > if $\varepsilon_2(\bar{S}) < \bar{\varepsilon}$, set $\bar{\varepsilon} = \varepsilon_2(U)$ and $U = \bar{S}$
> > end
> until the number of trials T exceeds N
> return U and $\bar{\varepsilon}$

Notice that once U (the set of inliers) has been found, as a final step a classical linear optimization may be applied (e.g., Least Mean Squares) to obtain the optimal estimation of F. In practice, this final step may be rather useless since the estimation of F performed by the algorithm (using sets of 7 points) is generally excellent (this will be confirmed by Section 12.6.1 and Figure 12.8).

If we assume that among n point matches, k are correct (i.e., we have $n - k$ outliers), then the probability of selecting at least one set of 7 correct point matches in N trials is

$$q = 1 - \left(1 - \frac{\binom{k}{7}}{\binom{n}{7}}\right)^N = 1 - \left(1 - \prod_{i=0}^{6} \frac{k-i}{n-i}\right)^N.$$

For $q = 95\%$ (i.e., $\mathrm{Log}_e(1-q) \simeq -3$), we can see that the number of trials required is approximately

$$N \simeq 3 \left(\frac{n}{k}\right)^7, \tag{12.12}$$

which allows a proportion $p = 1 - k/n$ of outliers of about 70% to stay in reasonable computation time limits (10,000 trials). A proportion of 90% outliers would require

$N = 30{,}000{,}000$ trials, which is not feasible. However, we will see that this estimate of N is questionable, and that a rigid set can be found among more than 90% outliers in much less than 30,000,000 trials.

12.5.3 Optimized Random Sampling Algorithm (ORSA)

There is a simple way to improve the stochastic algorithm. It relies on the idea that, on average, the proportion of outliers should be smaller than p among the most meaningful rigid sets. This suggests that the final optimization step below should be added at the end of the previous Random Sampling Algorithm (just before the "return U and $\bar{\varepsilon}$" line).

> set $\bar{\varepsilon} = \varepsilon_2(U)$
> repeat
>   generate a random set T of 7 point matches among U
>   for each fundamental matrix F associated to T
>   compute the most meaningful rigid set $\bar{S} = \bar{S}(F)$ associated to F
>   if $\varepsilon_2(\bar{S}) < \bar{\varepsilon}$, set $\bar{\varepsilon} = \varepsilon_2(\bar{S})$ and $U = \bar{S}$
>   end
> until the number of trials T exceeds N_{opt}

If U is absolutely meaningful (i.e., $\varepsilon_2(U) < 1$), this optimization step converges very quickly and N_{opt} does not need to be large. In any case, it cannot be harmful since $\bar{\varepsilon}$ is constrained to decrease. In practice set $N_{opt} = N/10$ and apply this optimization step to the first absolute meaningful set found by random sampling or if none has been found after $N - N_{opt}$ trials to the most meaningful set found so far. In this way the total number of trials cannot exceed N (but can be as small as $N_{opt} + 1$), while improving dramatically the detection of outliers, as shown in Section 12.6. We will refer to this algorithm bt the ORSA (Optimized Random Sampling Algorithm) acronym.

12.6 Experiments

12.6.1 Checking All Matchings

In two digital images (resolution 800×600) of the same scene, a set S of 23 point matches with a 1-pixel accuracy (see Figure 12.7) was manually recorded. Taking a snapshot of a regular grid with the same camera showed that geometric distortions were between 0 and 5 pixels. Thus, a rigidity coefficient between 0 and

$$\alpha = \frac{2 \cdot 5 \cdot \sqrt{800^2 + 600^2}}{800 \cdot 600} \simeq 10^{-1.68}$$

Fig. 12.7 The first stereo pair with 23 point matches and the best epipolar geometry recovered by the systematic exploration of all seven-point transforms.

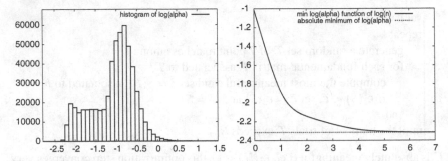

Fig. 12.8 Left: the histogram of $\text{Log}\,\alpha_F(S)$ for all possible F obtained from sets of seven-point matches taken out from S. From this histogram (measured at a much finer scale), the expectation of $\min_{i=1..n}\text{Log}\,\alpha_{F_i}(S)$ was comuted (the F_i's being obtained from independent random sets of seven-point matches). This expectation is represented as a function of $\text{Log}\,n$ on the right. We can see that the systematic exploration of all sets of seven-point matches is not really required: An excellent rigidity coefficient ($\text{Log}\,\alpha = -2.23$) could be obtained in about 1000 random trials, which is to be compared to the $550,000$ trials needed to obtain the best possible rigidity coefficient $\text{Log}\,\alpha = -2.31$ (all logarithms taken in base 10).

could be expected. Using the systematic combinatorial search yielded $\text{Log}(\alpha) = -2.31$ with the best seven-point transform, whereas the least mean squares optimization applied to the 23×9 linear system yielded a transform F such that $\text{Log}\,\alpha_F(S) = -2.32$ (note that the least mean squares optimization does not significantly improve the estimation of F, compared to the systematic combinatorial search). As expected, the meaningfulness of the rigidity test was very good:

$$\varepsilon_1 = 3 \cdot \binom{23}{7} 10^{-2.31(23-7)} \simeq 10^{-31}.$$

One could argue that testing $\binom{23}{7} \simeq 250,000$ configurations takes time (25 seconds), but in fact a nearly optimal accuracy is achieved by testing only a small number of random seven-point transforms, as shown in Figure 12.8. This justifies the systematic use of the random sampling algorithm we presented in Section 12.5.

12.6.2 Detecting Outliers

Outliers were simulated by replacing the first 10 point matches (among 23) by random points. Then for each fundamental matrix F that yielded by a seven-point transform, the most rigid set size k, $S_k(F)$ (see Section 12.5.1), as well as its rigidity $\alpha^k(F) = \alpha_F(S_k(F))$ and its meaningfulness $\varepsilon^k(F) = \varepsilon_2(\alpha_k(F), 23, k)$ were computed. In Figure 12.9 are represented the two functions $k \mapsto \bar{\alpha}(k) = \min_F \alpha^k(F)$ and $k \mapsto \bar{\varepsilon}(k) = \varepsilon^k(F)$. The function $\bar{\alpha}(k)$ is increasing with respect to k (by construction), and presents only a small gap between $k = 13$ and $k = 14$, from which it would be difficult (and uncertain) to directly detect the outliers. On the contrary, the function $\bar{\varepsilon}(k)$ presents a strong minimum for $k = 13$, which identifies without doubt the true set of 13 point matches and the 10 outliers. The epipolar geometry is well recovered too, as shown on Figure 12.10.

12.6.3 Evaluation of the Optimized Random Sampling Algorithm

In some situations, the number of point matches n may be large (say, $n \geq 25$) and a systematic exploration of all sets of seven-point matches is not possible (see Table 12.1). In this case the Optimized Random Sampling Algorithm (ORSA) presented in Section 12.5.3 can be used. It is interesting to use this algorithm even for small values of n, as shown in Figure 12.8. In this subsection, we propose a systematic evaluation of ORSA and we prove in particular the usefulness of the final optimization step introduced in Section 12.5.3.

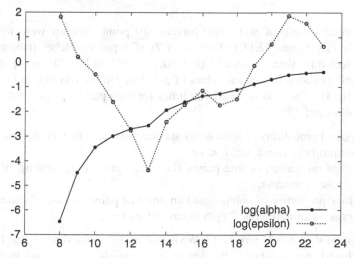

Fig. 12.9 Rigidity error $\bar{\alpha}(k)$ and meaningfulness $\bar{\varepsilon}(k)$ of the most rigid subset of size k (log scale). Whereas $\bar{\alpha}(k)$ can hardly be used alone to detect outliers, $\bar{\varepsilon}(k)$ presents a strong minimum which permits us to detect the 13 inliers easily.

Fig. 12.10 When we replace the first 10 point matches (among 23) by random points, the use of the ε_2 criterion allows us to identify the meaningfully rigid set of 13 inliers and to recover the right epipolar lines (shown in white).

Fig. 12.11 The second stereo pair with 70 point matches (only 30 are shown here for clarity). The epipolar geometry is recovered in no time and yields $\text{Log}\,\alpha = -1.92$ and $\text{Log}\,\varepsilon = -110$.

In a second couple of 800×600 images, 70 point matches were manually recorded by (see Figure 12.11.) Then a set T_k of k point matches (among these 70) was selected for three values of k ($k = 10$, $k = 30$, and $k = 70$) and $kp/(1-p)$ random outliers added for several values of p before ORSA was run on these data with $N = 10,000$. This was repeated 100 times for each pair (k, p) in order to compute the following:

- the empirical probability of success (by success, we mean that an absolute meaningful set has been found, i.e., $\bar{\varepsilon} < 1$);
- the average proportion of true points (i.e., belonging to T_k) among all points found in case of success;
- the average proportion of points found among true points in case of success;
- the average meaningfulness $< \varepsilon_2 >$ in case of success.

The results are shown in Figure 12.12. We notice in these experiments that for $N = 10,000$ (which is not very large), the detection thresholds are good and, in the case of success, the recovery of the true set of inliers is almost perfect (notice that by adding random points, we may also add new pairs that match well the true epipolar

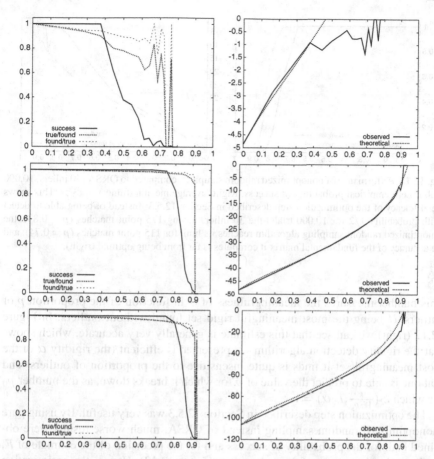

Fig. 12.12 Performance of the Optimized Random Sampling Algorithm (10,000 trials) in function of the proportion p of outliers (horizontal axis). Top row: 10 inliers; middle row: 30 inliers; bottom row: 70 inliers. Left: empirical probability of success (detection of an absolutely meaningful rigid set), average proportion of true inliers among the most rigid set found (in case of success), and average proportion of points found among true inliers (still in case of success). Right: observed and theoretical average meaningfulness $< \varepsilon_2 >$ in the case of success.

transform). With a 50% probability of success (which would become $1 - 0.5^5 = 97\%$ for $N = 50,000$), we can allow a proportion of outliers equal to $p = 0.5$ for $n = 10$, to $p = 0.83$ for $n = 30$, and to $p = 0.86$ for $n = 70$. This last case means that 70 inliers still are detectable among 500 points and proves that (12.12) underestimates by far the performance of ORSA.

Another interesting point is that the ε_2 criterion allows us to predict the algorithm performance. From the rigidity coefficient α expected for inliers (and measured before adding outliers), one can predict the average meaningfulness $< \varepsilon_2 >$ by using the theoretical value

$$\varepsilon_2 \left(\frac{k}{1-p}, k, \alpha \right).$$

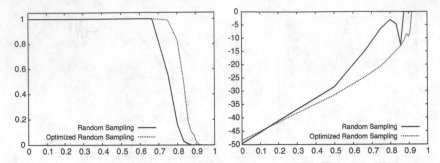

Fig. 12.13 Performance of nonoptimized random sampling, compared to ORSA (30 inliers, 10,000 trials). Left: empirical probability of success. Right: average meaningfulness $< \varepsilon_2 >$. This shows the efficiency of the optimization step described in Section 12.5.3: Instead of being able to detect (with probability 1/2 for 10,000 trials) the 30 inliers among 175 point matches ($p = 0.83$), the nonoptimized random sampling algorithm reaches its limit for 115 point matches ($p = 0.74$), and the accuracy of the fundamental matrix it computes is far from being optimal (right).

This number measures the meaningfulness of U in function of the proportion p of outliers, U being the most meaningful rigid set found without outliers. In Figure 12.12 (right) we can see that this estimate is generally very accurate, which shows that the rigidity detection algorithm we presented is efficient (the rigidity α of the most meaningful set it finds is quite insensitive to the proportion of outliers) and that one is able to predict the value of p for which it breaks down, as the number p_0 for which $\varepsilon_2(\frac{k}{1-p_0}, k, \alpha) = 1$.

The optimization step described in Section 12.5.3 was very useful. By using pure (nonoptimized) random sampling instead of ORSA, much worse results were obtained both for the probability of success and for the precision of the estimate of F, as shown by the meaningfulness loss (see Figure 12.13). Hence, the optimization process not only speeds up the algorithm, but it often lets us find a meaningful rigid set when none has been found in $N - N_{opt}$ trials by choosing sets of seven-point matches only from the most meaningful rigid set found so far.

12.7 Bibliographic Notes

12.7.1 Stereovision

Stereovision and, more generally, structure from motion are based on the fact that two images of a static scene taken by a pinhole camera from two different viewpoints are redundant. Two methods use this redundancy to recover the 3-D structure. The first uses dense matching maps (optical flow). The second is based on feature matching –that is the correspondence of several physical parts of the scene (typically corners or lines) in the two images. The analysis of discrete point matches allows us to recover, up to a scale factor, not only the 3-D position of the points before their projection in each image but also the camera motion and its intrinsic parameters [LH81, FLT87, Har97a]. All of these methods use the fact that two perspective projections of a static scene are related through the fundamental matrix. This highlights the importance of estimating the fundamental matrix from point matches and explains the large attention it has received in the last two decades [FL01, LF96, OG01, SAP01, TM97, ZDFL01, Zha98].

12.7.2 Estimating the Fundamental Matrix from Point Matches

For cameras with calibrated internal parameters, only the essential matrix has to be recovered and the epipolar geometry depends on 5 parameters: 2 for the 3-D translation (the unknown scale factor lets us impose $\|T\| = 1$ for example), and 3 for the 3-D rotation (matrix R). Since each point correspondence produces one equation like (12.1), a minimum of 5 matchings is required to recover the epipolar geometry. As shown by Faugeras [FL01], 5 matchings are sufficient, but the computation is rather complex and unstable and provides in general 10 solutions.

For this reason, and because more points are needed in the case of cameras with uncalibrated internal parameters, the much simpler linear eight-point algorithm is preferred in general. Introduced by Longuet-Higgins [LH81], this algorithm has been refined in several ways. The classical formulation is the following. For each point match (m_i, m_i'), Equation (12.1) can be rewritten as a linear equation depending on the 9 coefficients of F, viewed as a 9-dimensional vector f. Hence 8 point correspondences yield a 8×9 linear system

$$Af = 0,$$

where the i-th line of A is obtained from the coordinates of m_i and m_i' (as noticed by Hartley [Har97b], these coordinates must be normalized to improve the conditioning of the matrix A). Since f is defined up to a scale factor, this system lets us recover f. This can be done by looking for a unit norm vector in the (generally one-dimensional) null space of A. This method has a small drawback. The solution F obtained is not a rank 2 matrix in general, which means that the "epipolar geometry" underlying F may not be consistent with a physical realization (epipolar lines may not all intersect at the epipole). This problem can be solved by using the singular value decomposition of F and forcing the smallest eigenvalue to 0, or with some kind of parallax estimation (see [BM95]).

The seven-point algorithm, used repeatedly throughout this chapter, avoids this issue by considering only 7 point matches while enforcing the rank 2 constraint by solving a third degree polynomial equation. Each of the 3 possible fundamental matrices obtained with this method defines a true epipolar geometry, where all epipolar lines pass through a unique point (the epipole) while each of the 7 point correspondences is realized exactly.

When dealing with more than 7 or 8 points several methods exist. The classical Least Square Minimization (LMS) looks for F that minimizes $\sum_i (m_i'^T F m_i)^2$. The solution, easily computed using linear algebra, can then be improved by minimizing a more geometric criterion [LF96], the sum of the squared distances from each point to the corresponding epipolar line, $\sum_i r_i^2$, where $r_i = \text{dist}(m_i', F m_i)$. This criterion can be symmetrized into $\sum_i r_i^2 + r_i'^2$, with $r_i' = \text{dist}(m_i, F^T m_i')$. These methods and similar ones heavily suffer from badly located points or false matches. This is why robust methods have been introduced to reject outliers in the optimization process.

12.7.3 Robust Methods

Among the most efficient robust methods are M-estimators, LMedS, RANSAC [SAP01, TM97], and the recent Tensor Voting [TML01]. M-estimators (see [ZDFL94] for example) try to reduce the influence of outliers by applying an appropriate weighting of the distances. The criterion to minimize becomes $\sum_i \rho(r_i)$, where ρ grows more slowly than the square function. As pointed out by Faugeras and Luong, *it seems difficult to select a ρ-function for general use without being rather arbitrary* [FL01]. RANSAC methods use a set of fundamental matrices obtained from random sets of seven-point matches (with the seven-point algorithm) to find the largest consensus set –that is, the largest set of point matches achieving a given match accuracy. Like M-estimators, RANSAC suffers from arbitrary choice since an accuracy threshold has to be preset. The Tensor Voting method, which relies on the scalar selection of a 8-dimensional hyperplane, suffers from the same kind of problem.

12.7.4 Binocular Grouping

The a-contrario model presented in this chapter was developed by Moisan and Stival in [MS04]. It yields absolute thresholds for rigidity detection and allows the design of an optimized random sampling algorithm that is able to estimate the epipolar geometry between two images even for a very large proportion of outliers.

Except in the case of a small number (n) of point matches (say, $n \leq 25$) where a deterministic algorithm (systematic search) may be used, its implementation relies on a stochastic algorithm following the random sampling consensus (RANSAC) introduced in the context of image analysis by Fischler and Bolles [FB81]. The rigidity detection criterion it proposes dramatically improves the performance of classical RANSAC algorithms. Using this approach, one can detect rigidity and provide a good estimate of the fundamental matrix when the initial set of point matches contains up to 90% of outliers. This outperforms the best currently known algorithms such as M-estimators, LMedS, and RANSAC, which typically break down around 50% of outliers. As far as we know, even the recent Tensor Voting technique has not been reported to work with more than 65% of outliers.

Why does this approach outperform classical random sampling algorithms? In the RANSAC algorithm, a threshold has to be preset arbitrarily to decide whether a set of point matches is "compatible" with a fundamental matrix F. Then, using random sampling, the largest compatible set is selected. The algorithm presented in this chapter works in a similar way, except that no arbitrary threshold has to be selected: The ε_2 criterion is used to rate the meaningfulness of a set of point matches in function of its size (k) and its rigidity (α). Even better, it tells in the end if the most meaningful set found by the algorithm can be considered as meaningfully rigid (i.e., $\varepsilon_2 \ll 1$) or if it could have been that rigid by accident.

12.7.5 Applications of Binocular Grouping

The model of binocular grouping presented in this chapter can be applied to stereo pairs or in the context of Structure From Motion since no assumption is required on the magnitude of the camera motion. It could also be used to detect several rigid motions in an image sequence (see [TZM95] for an example), although no experiment has been made for that application. The case of nonmatched points reveals the existence of an absolute accuracy threshold (1/32) in the matching process and of an optimal value for the number of points to be matched depending on the expected accuracy and on the expected proportion of outliers. It suggests that additional measures should be used for point correspondence, like color or other perspective-invariant qualities. The meaningful rigidity criterion does not measure the uncertainty of the fundamental matrix, and it might happen that a very meaningful rigid set is found with a very uncertain associated fundamental matrix. Hence, computing this uncertainty (see [Zha98] for a review) may still be useful for practical applications. The approach we presented could also be applied to simpler motions, as mentioned in Section 12.4.4. The case of line matches [FLT87] or of simultaneous points and line matches could probably be treated as well. The extension to more than two views and to the detection of several simultaneous rigid motions could also have interesting applications.

12.8 Exercise

12.8.1 Epipolar Geometry

Let us describe a calibrated pinhole camera by (O, i, j, k, f), where $O \in \mathbb{R}^3$ (optical center), (i, j, k) is an orthonormal basis of $\mathbb{R}^3$, and f is a positive number (focal length). Given a point M of the space such that $\overrightarrow{OM}.k > f$ (i.e., M is a visible point), the optical ray $[MO)$ intersects the image plane $\{P; \overrightarrow{OP}.k = f\}$ at a unique point P.

Let (O', i', j', k', f') be a second calibrated pinhole camera (we assume that $O' \neq O$). If M is visible by this camera, the optical ray $[MO')$ intersects the second image plane at a point P' such that $\overrightarrow{O'P'}.k' = f'$.

1) Show that

$$\overrightarrow{O'P'}.(\overrightarrow{OO'} \wedge \overrightarrow{OP}) = 0. \tag{12.13}$$

2) We write $\overrightarrow{OO'} = t_x i + t_y j + t_z k$, and define

$$T = \begin{pmatrix} 0 & -t_z & t_y \\ t_z & 0 & -t_x \\ -t_y & t_x & 0 \end{pmatrix}.$$

Show that T represents the linear mapping $v \mapsto \overrightarrow{OO'} \wedge v$ in the basis (i, j, k).

3) Show that if $\overrightarrow{OP} = Xi + Yj + fk$ and $\overrightarrow{O'P'} = X'i' + Y'j' + f'k'$, then there exists a 3×3 rotation matrix R such that

$$(X' \, Y' \, f') \, E \, (X \, Y \, f)^T = 0 \quad \text{with} \quad E = RT. \tag{12.14}$$

The matrix E is called the *essential matrix*.

4) Now we suppose that the two cameras are not calibrated. This means that instead of X and Y, we only observe $x = \alpha X + x_0$ and $y = \beta Y + y_0$ for some positive α and β (as well, $x' = \alpha' X' + x'_0$ and $y' = \beta' Y' + y'_0$). Show that there exist two invertible triangular matrices C an C' (called *calibration matrices*) such that

$$(x' \, y' \, 1) \, F \, (x \, y \, 1)^T = 0 \quad \text{with} \quad F = C'^{-T} E C^{-1}. \tag{12.15}$$

The matrix F is called the *fundamental matrix* and (12.15) is the *epipolar constraint* mentionned in Section 12.2.1 (Equation (12.1)).

5) Show that E and F both have rank 2 and that the two nonzero eigenvalues of $E^T E$ are equal.

Chapter 13
A Psychophysical Study of the Helmholtz Principle

13.1 Introduction

Is our perception driven by the Helmholtz principle? In this chapter, we describe two psycho-visual setups. The experiments were designed to test the ability of a subject to detect the presence of certain gestalts in an image. The first psycho-visual experiment tests human ability to detect dark or bright squares in a white noise image (Section 13.2). The second experiment tests alignment perception (Section 13.3).

For each experiment, the responses of the tested subjects were measured as a function of two parameters, namely the size of the gestalt to be detected and the amount of noise. As we will see, the match between perception thresholds and theoretical ones predicted by the Helmholtz principle is good. This is checked by comparing the theoretical iso-meaningfulness curves in parameter space with the observed ones.

13.2 Detection of Squares

13.2.1 Protocol

The first experiment deals with the detection of a square in a digital synthetic image. The $N \times N$ black-and-white image is the realization of a Bernoulli process with parameter $p(x,y)$. Each pixel gray level $u(x,y)$ is chosen randomly and independently to be black (0) with probability $p(x,y)$ and white (255) with probability $1 - p(x,y)$. The parameter $p(x,y)$ takes the constant value $\bar{d} \in [0,1]$ inside a random square domain of the image and the value $1/2$ outside this square. The square location, the side length l, and the average density $\bar{d}$ are chosen randomly, uniformly and independently. The resulting image is a binary white noise image containing a clear ($\bar{d} < 1/2$) or a dark ($\bar{d} > 1/2$) square that may be visible or not according to its size and the choice of $\bar{d}$. Such an image is shown in Figure 13.1.

Fig. 13.1 Example of the test image used for square detection.

A sequence of 100 such random images was presented to each subject. Each image appeared on the screen for 1.5 seconds. For each image, the subject had to answer to the question "Can you see a square in this image?" If his answer was positive, he had to press a yes key in the 1.5-second time lapse of image display. Between each image, a blank image was displayed during approximately 0.5 second to avoid interferences between successive images.

For each image, the yes-no answer was recorded along with the square side length (l) and its observed relative density $\delta = |d - 1/2|$, where d is the ratio of white pixels contained in the square. Note that the expectation of d is $\bar{d}$ but that both numbers may slightly differ. To each image was associated a point in the (δ, l) plane. Thus, each subject's answers could be displayed by two clouds of dots (the "yes" cloud and the "no" cloud) in the (δ, l) plane (see Figure 13.3).

13.2.2 Prediction

If perception is based on the Helmholtz principle, the square detection is made *a contrario* against the hypothesis that the image is a white noise –in the present case, a Bernoulli noise with parameter $1/2$. The number of false alarms (NFA) associated to a square with side length l and relative density δ is

$$\text{NFA}(l, \delta) = N^3 \cdot \mathbb{P}\left[\left|\frac{S_{l^2}}{l^2} - \frac{1}{2}\right| \geq \delta\right], \tag{13.1}$$

where S_n denotes the sum of n independent Bernoulli random variables with parameter $1/2$. This implies that

$$\mathbb{P}[S_n \geq k] = 2^{-n} \sum_{j=k}^{n} \binom{n}{k}.$$

The first factor of (13.1), N^3, counts the number of possible squares (N^2 locations and N side lengths). With the usual notations, we have

$$\mathbb{P}\left[\left|\frac{S_{l^2}}{l^2}-\frac{1}{2}\right|\geq\delta\right]=\mathbb{P}\left[S_{l^2}\geq\frac{l^2}{2}+l^2\delta\right]+\mathbb{P}\left[S_{l^2}\leq\frac{l^2}{2}-l^2\delta\right]$$

$$=2B\left(l^2,\frac{l^2}{2}+l^2\delta,\frac{1}{2}\right),\qquad\qquad\text{(13.2)}$$

so that a large deviation estimate (see Proposition 6) yields

$$\log\mathrm{NFA}(l,\delta)=3\log N-\frac{l^2}{2}\left((1+2\delta)\log(1+2\delta)+(1-2\delta)\log(1-2\delta)\right)$$

$$+\log 2+\underset{l\to+\infty}{o}(l^2).\qquad\qquad\text{(13.3)}$$

Each level line defined by $\mathrm{NFA}(l,\delta)=\varepsilon$ separates two regions in the (δ,l) plane: the squares that we are ε-meaningful and the other ones. For l large enough, the equation of these lines is well approximated by

$$l\simeq\sqrt{\frac{2(3\log N+\log 2-\log\varepsilon)}{(1+2\delta)\log(1+2\delta)+(1-2\delta)\log(1-2\delta)}}.\qquad\qquad\text{(13.4)}$$

If, in addition, δ is small enough, since

$$(1+2\delta)\log(1+2\delta)+(1-2\delta)\log(1-2\delta)=4\delta^2+O(\delta^4),\qquad\text{(13.5)}$$

(13.4) may be simplified into

$$\delta\cdot l=\sqrt{\frac{3\log N+\log 2-\log\varepsilon}{2}}.\qquad\qquad\text{(13.6)}$$

Hence, the separation curves look like hyperbolas. The curves obtained by (13.4) are represented in Figure 13.2 for several values of ε. If our visual perception has

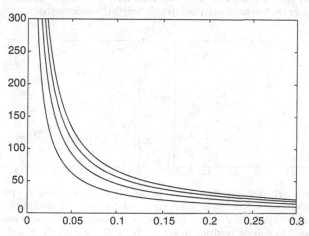

Fig. 13.2 Thresholds in the (δ,l) plane (square density and side length) predicted by the Helmholtz principle for different values of ε ($\varepsilon=10^{-0},10^{-10},10^{-20},10^{-30}$) when $N=600$.

something to do with the Helmholtz principle, the experimental perception thresholds should correspond to one of these NFA level lines. The level ε of this line would indicate a biological confidence parameter.

13.2.3 Results

Eight persons participated in the experiment described. Each subject was submitted a first (nonrecorded) training set of 50 images, then a real set of 100 images. No other explanation was given to the subjects than just the written question "can you see..." Each image had size 600×600 and was displayed on a 15-inch 1600×1200 LCD panel. The answers are reported in Figure 13.3. Note that the values chosen randomly for $\bar{d}$ and l excluded the domain $\{(\delta, l), \ \delta > 0.15 \text{ and } l > 60\}$, for which the detection of the square is too obvious.

The collected data fit well to the Helmholtz model if $l \le 100$ and yield $\varepsilon \simeq 10^{-13}$ (see Figure 13.4). Yet, for larger values of l, the measured perception threshold differs from the one predicted by the Helmholtz principle (see Figure 13.5).

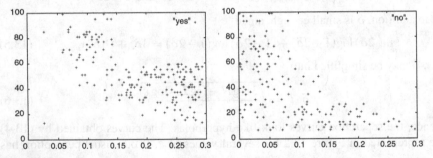

Fig. 13.3 Positive (left) and negative (right) answers for $l \le 100$, as a function of the relative density of the square (δ, horizontal axis) and its side length (l, vertical axis).

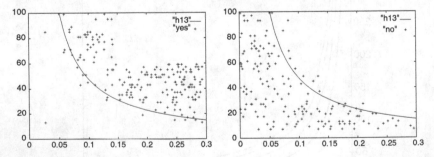

Fig. 13.4 Positive (left) and negative (right) answers for $l \le 100$ and the prediction curve ($\varepsilon = 10^{-13}$). The qualitative fit to the model is good, but the value of ε unrealistic. It is not likely that a pixel counting be performed in such fine images. A multiscale process is probably at work. This can be avoided by making synthetic images with much larger details, as will be done in the next experiment.

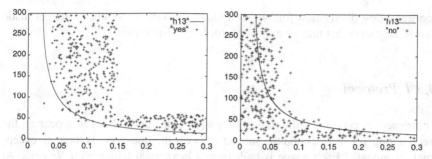

Fig. 13.5 Comparison between the prediction curve, positive (left) and negative (right) answers for $l \leq 300$. There is no good fit for large l's. This seems due to an error in experiment design. The images and features used are too large with too many small pixels. Multiscale processes are at work in the perception of large squares. It is not likely that all pixels inside the square are counted. Their detection is probably made by a contrast perception on their boundary and does not involve l^2 pixels, as assumed in the theoretical model.

13.2.4 Discussion

Since the experiments involve two parameters (δ and l) whereas the model only has one (ε), the fit obtained between the measured data for $l \leq 100$ and the model seems to be relevant. How can one explain the lack of fit for $l \geq 100$? For such a large square, our visual system has to zoom out the image. Thus, a model based on the fact that inside the square, each pixel (black or white) is taken into account becomes questionable. It is likely that the square perception involves its boundary points, not its interior points. Thus, one can conjecture that the images in this experiment have too many pixels. Another argument against our perception taking into account each pixel in such large images is the following. The value for the observed perception threshold ε is too small and not realistic. Indeed, why should a subject allow for so few false alarms? The next experiment will better address this sampling issue by making images with much larger details. Then ε goes up to 10^{-2}.

Should the responses of several subjects be fused? Performing meany measurements with one single subject is difficult. A loss of concentration cannot be avoided after a certain number of experiments. Taking several subjects could have made the separation between "yes" and "no" answers more fuzzy. The threshold that each subject chose might have depended on the interpretation of the question. "Do I have to say yes when I am sure that I see a square or when I think that there may be a square?" This question was actually asked by some subject,s but no directions were given. As a matter of fact, no blur in the separation curve was observed by mixing the subjects answers.

13.3 Detection of Alignments

To avoid the resolution issue mentioned in the above experiment, an experiment involving a small number of objects was designed in the classic gestaltist style. We refer to the books of Metzger [Met75] and Kanizsa [Kan97, Kan91, Kan79]. The experiment described below follows their ideas, by showing synthetic images made of small segments that may or may not form large alignments.

13.3.1 Protocol

The protocol is essentially the same as for square detection. Images appear on the screen during 1.5 seconds and the subject answers the question "Is there an exceptional alignment?" Each image is built from a hexagonal lattice with N^2 cells. At the center of each cell, there may be a small segment or not. This small segment has three possible orientations: 0, 120, or 240 degrees, so that each segment points toward the centers of two opposite neighboring cells.

To create each image, an alignment is generated in the following way. A position, a length l, and an orientation are randomly chosen. They determine a unique segment in the lattice made of l aligned small segments. These segments are constrained to have the same direction as the alignment itself. Depending on its length, such a segment can be conspicuous or not. Indeed, it is endowed in a clutter of random segments. Let us call background cells the cells not belonging to the alignment. To generate this background clutter for each image, a density $\bar{d}$ is randomly and uniformly chosen in $[0, 1/2]$. Each background cell is empty with probability $1 - d$ and contains a uniform random choice of the three possible segments with probability $d/3$. All random variables are independent. Such a random image is shown in Figure 13.6.

For each image, the answer of the subject is recorded (yes or no), along with the alignment length (l) and the density (d) of the background little segments (as earlier, the expectation of d is $\bar{d}$). Each image is then represented by a point in the (d, l) plane.

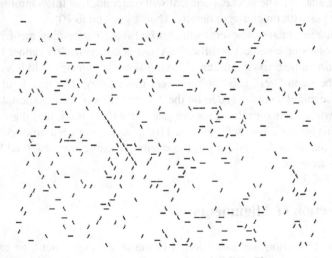

Fig. 13.6 Example of test image used for alignment detection ($N = 50$).

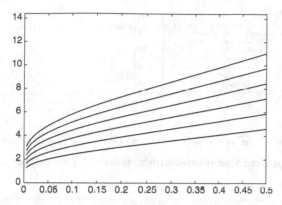

Fig. 13.7 Thresholds in the (d,l) plane (segment density and alignment length) predicted by the Helmholtz principle for different values of ε ($\varepsilon = 10^{-0}, 10^{-1}, 10^{-2}, \ldots, 10^{-5}$) when $N = 50$.

13.3.2 Prediction

Applying the Helmholtz principle to this experiment, the number of false alarms associated with the existence of a length l segment alignment is in the background model is

$$\text{NFA} = \frac{3}{2}N^3 \left(\frac{d}{3}\right)^l. \tag{13.7}$$

The number $3N^3/2$ approximately counts the number of possible alignments on the image (three orientations, N^2 positions for the center, $N/2$ possibilities for the segment length). The second term is simply the probability that the l cells of the alignment have the proper orientation, knowing the empirical density d of the background small segments. From (13.7), we deduce that the threshold curve in the (d,l) plane corresponding to NFA $\leq \varepsilon$ is defined by

$$l = \frac{C}{\log(d/3)}, \qquad \text{where} \quad C = \log(\varepsilon) - \log\left(\frac{3}{2}N^3\right). \tag{13.8}$$

Some of these curves are displayed in Figure 13.7.

13.3.3 Results

Seven persons participated in this experiment, yielding 900 answers. As in the square detection, each subject was first submitted to a shorter (nonrecorded) training set, needed by the subject to understand the procedure well. The answers are shown in Figure 13.8.

In Figure 13.9, one can check that that for $\varepsilon \simeq 10^{-2}$, the threshold curve predicted by the Helmholtz principle separates well the "yes" and "no" answers. The fit can be

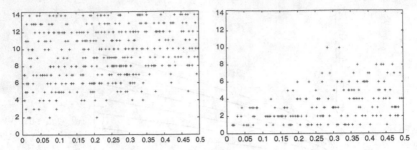

Fig. 13.8 Positive (left) and negative answers (right) in the (d,l) plane (segment density and alignment length).

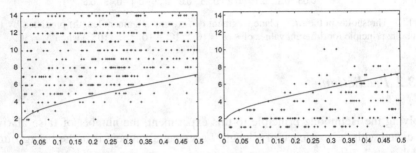

Fig. 13.9 Positive (left) and negative answers (right) and the prediction curve ($\varepsilon = 10^{-2}$).

hardly improved since the answers are somewhat mixed up (the threshold is fuzzy). The experiment design seems to lead to a realistic perception threshold. The subjects allowed themselves 1 false detection in 100 experiments.

13.4 Conclusion

The experiments have proven that the threshold curves predicted by the Helmholtz principle fit qualitatively human perception. The observed quantitative perception threshold seems correct in the second experiment and unrealistic in the first one. This indicates that a tight control of the visual sampling must be performed in experiment design. In that way it might be possible to find absolute psychophysical detection thresholds –in other words, to predict which NFA humans allow themselves. Such an absolute detection threshold ($\varepsilon = 10^{-2}$) seems to have been attained in the second experiment.

The square detection experiment was not satisfactorily designed. In this experiment the images were too large and the pixels too small. In the perception of such a large image, we know from gestaltists and neurophysiology that masking processes and multiscale analysis processes are at work. This may explain why the best fit for the square detection corresponds to $\varepsilon = 10^{-13}$, which is not a realistic NFA at all.

A significant improvement in the method could be realized by designing several experiments from which only one single threshold curve is predicted (corresponding to a kind of "universal" value of ε), instead of a family of curves from which the best fit is selected, as above.

13.5 Bibliographic Notes

The experiments presented in this chapter were published in [DMM03a]. They follow typical setups of visual psychophysics: A shape is put in an image and noise is added before the perception threshold is measured as a function of the amount of noise and of the shape variability (size, contrast, etc.). There are several statistical theories that may be used as a reference to rate the ability of our visual system. Among them, the theory of the Ideal Observer (derived from Bayesian decision theory) is commonly used since it yields an optimal performance bound for any observer. However, it requires perfect knowledge of the image statistics and a model of the detected pattern. This may explain why human perception generally performs poorly compared to the Ideal Observer. To give an example, Pelly, Farell, and Moore [PFM03] recently proved that our ability to read words in a noisy image is far from reaching the performance of an Ideal Observer based on the matching of an image template of each word. This fact is easily explained. We do not recognize a word as a whole precise image. We do not have in mind an accurate letter model but a qualitative one. Thus, the experiment was not well designed, in about the same way our above square experiment was not. The same lack of performance has been studied by Legge, Kersten, and Burgess in the case of contrast discrimination to explain why "even the highest contrast sensitivities that humans can achieve for the detection of targets on uniform fields fall far short of ideal values" [LKB87].

Chapter 14
Back to the Gestalt Programme

Chapter 2 proposed a classification of gestalt grouping processes. The basic grouping laws that group points if they share some geometric quality were called partial gestalts. A synopsis of all partial gestalts computed in this book is presented in Section 14.1. At the end of the chapter, a review of other gestalts computed in recent works confirms that all partial gestalts can be computed by similar methods. Thus, the focus in the computational gestalt programme should now be to formalize the more general gestalt principles, starting with gestalt collaboration and competition. How far do we stand in that direction? One of the grouping laws, the vanishing point detector, uses previously calculated gestalts, namely alignments. Thus, at least in this case, the recursive gestalt building up has been addressed. In Section 14.2 a successful complete gestalt analysis will be performed on a digital image by combining most partial gestalts computed so far. On the dark side, Section 14.3 presents a series of striking experimental examples showing perceptually wrong detections in the absence of collateral inhibition between conflicting partial gestalts. These experiments show that the gestalt programme still lacks some fundamental principle on gestalt interaction. A good candidate would be an extension of the exclusion principle introduced in Chapter 6. Another possibility would be to go back to variational principles like the Minimum Description Length principle. We address this possibility in the next and last chapter.

14.1 Partial Gestalts Computed So Far

Table 14.1 summarizes the very similar formulas computing the number of false alarms (NFA) of the following partial gestalts: *alignments (of orientations in a digital image), contrasted boundaries, similarities for some quality measured by a real number (gray level, orientation, etc.), vanishing points*, and the basic *vicinity* gestalt. Details of the computation of these partial gestalts were given in Chapters 3, 5, 7, 8, 9, and 11. The *perspective binocular grouping* gestalt was omitted from Table 14.1 as it cannot be summarized in three lines. However, it is similar to *dot alignment*.

Table 14.1 Synopsis of partial gestalts, their parameters, and their NFA

GROUP LOOKED FOR	MEASUREMENTS	NUMBER OF FALSE ALARMS
Alignment of directions on a segment [DMM00]		
A discrete segment with points at Nyquist distance (i.e., 2)	k: number of aligned points l: number of points on the segment	$N_{segments} \cdot \mathcal{B}(l,k,p)$ $N_{segments} = N^4$ N^2: number of pixels in the image $p = 1/16$ (angular precision)
Contrasted edges and boundaries [DMM01b]		
A level line (or a piece of) with points at Nyquist distance (i.e., 2)	μ: minimum contrast (gradient norm) along the curve l: length of the curve	$N_{level\ lines} \cdot H(\mu)^l$ H: empirical cumulative distribution of the gradient norm on the image
Similarity of a uniform scalar quality (gray level, orientation, etc.) [DMM03b]		
A group of objects having a scalar quality q such that $a \leq q \leq b$	k: number of points in the group M: total number of objects	$\dfrac{L(L+1)}{2} \cdot \mathcal{B}\left(M,k,\dfrac{b-a+1}{L}\right)$ L: number of values ($q \in \{1,...,L\}$)
Similarity of a scalar quality with decreasing distribution (area, length, etc.) [DMM03b]		
A group of objects having a scalar quality q such that $a \leq q \leq b$	k: number of points in the group M: total number of objects	$\dfrac{L(L+1)}{2} \cdot \max_{p\in\mathcal{D}} \mathcal{B}\left(M,k,\sum_{i=a}^{b}p(i)\right)$ L: number of values ($q \in \{1,...,L\}$) $\mathcal{D}$: set of decreasing distributions on $\{1,...,L\}$

Alignment of points (or objects) [DMM03b]		
A group of points falling in a strip (region enclosed by two parallel lines)	$N_{strips} \cdot B(M,k,p)$	p: relative area of the strip The strips are quantized in position, width, and orientation	M: total number of points k: number of points falling in the strip

Vicinity: clusters of points (or objects) [DMM01a]		
A group of points falling in a region enclosed by a low-resolution curve	$N_{regions} \cdot \sum_{i=k}^{M} \binom{M}{i} \sigma^i (1-\sigma-\sigma')^{M-i}$	σ: relative area of the region σ': relative area of the thick low-resolution curve k: number of points falling in the region	M: total number of points $N_{regions} = N^2 qr2^L$: the low-resolution curves are quantized in resolution (q), thickness (r), location (N), and bounded in length (L).

Vanishing point: group of concurrent straight lines [ADV03]		
A group of straight lines meeting all in the same "vanishing region"	$M \cdot B(N,k,p_j)$	p_j: probability that a line meets the vanishing region. k: number of lines meeting the region	M: total number of vanishing regions N: number of lines detected in the image.

The first row of Table 14.1 treats the alignments in a digital image. A segment in the image containing enough points with orientations aligned with the segment is perceived as an alignment. The alignment gestalt was detailed and discussed thoroughly. Indeed, it can be extended to the detection of any other shape such as circles, conics, and actually any pattern of interest with fixed shape – for instance, a logo. The extension is trivial provided the shape to be detected is described by its orientation at each one of its l points. Then the decision of whether an observed shape matches a given query boils down to counting the number of points k where both orientations coincide. The number of false alarms in the detection is $\text{NFA} = M \cdot \mathcal{B}(l, k, p)$, where M is the number of possible poses of the pattern and p is the orientation precision. As an example showing that this extension works without further discussion, an *arc of circle* detector is tested in Section 14.3.2.

All of the other gestalts are built in a similar manner and we need not comment on them all. We do, however, notice that each one of them has required different probabilistic modeling for the Helhmholtz a-contrario model. In one instance, the edge or boundary detection, the a-contrario model is learned from the image. This opens a Pandora's box of possible uses of learned a-contrario models for every single gestalt quality. For instance, the alignment detector can use an observed distribution of orientations instead of the uniform distribution. This yields similar results, but the vertical and horizontal alignments are less favored. Indeed, the probability of getting these orientations in most images is higher than average.

The third and fourth rows in the table summarize the similarity gestalt: objects grouped by orientation, gray level, or any perceptually relevant scalar quality. The hope is that Table 14.1 does not need to be extended to infinity and that one already has a tool and enough variants to deal with any other gestalt problem.

14.2 Study of an Example

The main challenge is now to make partial gestalts interact and to answer the two main questions that stopped gestaltism, the resolution of gestalt *conflicts* and the principles of their *collaboration*. This is a major problem directly related to the so-called *binding and inhibition* problem in neurophysiology. Computer Vision seems to be the right research field to test abstract principles governing the collaboration of different gestalt qualities. A third related question is the search for principles governing the bottom-up building of gestalts. Thus, it seems interesting to do here some preliminary experiments combining several of the gestalt qualities that can be computed so far.

Let us undertake a complete study of a simple but real digital image involving almost all computational gestalts of Table 14.1. The analyzed image in Figure 14.1 is a common digital image – a scan of a photograph – and has blur and noise. The seeable objects are electrophoresis spots, which all have similar but varying shapes and colors and present some striking alignments. Actually, all of these perceptual remarks can be recovered in a fully automatic way by combining several partial gestalt grouping laws (Figure 14.2).

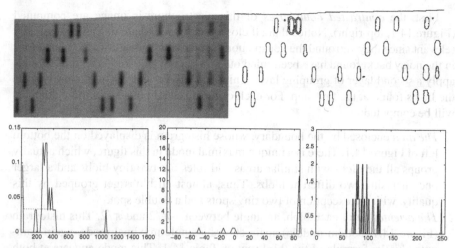

Fig. 14.1 Collaboration of gestalts. The objects tend to be grouped similarly by several different partial gestalts. First row: original DNA image (left) and its maximal meaningful boundaries (right). Second row, left: histogram of areas of the meaningful blobs. There is a unique maximal mode (256-416). The outliers are the double blob, the white background region, and the three tiny blobs. Second row, middle: histogram of orientations of the meaningful blobs (computed as the principal axis of each blob). There is a single maximal meaningful mode (interval). This mode is the interval 85-95 degrees. It contains 28 objects out of 32. The outliers are the white background region and three tiny spots. Second row, right: histogram of the mean gray levels inside each blob. There is a single maximal mode containing 30 objects out of 32, in the gray-level interval 74-130. The outliers are the background white region and the darkest spot.

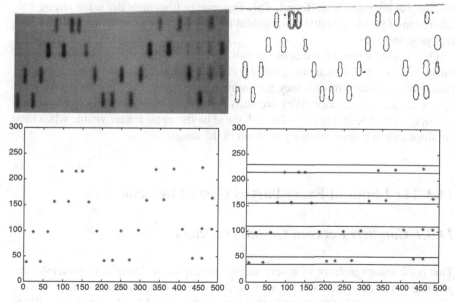

Fig. 14.2 Gestalt grouping principles at work for building an "order 3" gestalt (alignment of blobs of the same size). First row: original DNA image (left) and its maximal meaningful boundaries (right). Second row: left, barycenters of all meaningful regions whose area is inside the only maximal meaningful mode of the histogram of areas; right, meaningful alignments of these points.

First, the *contrasted boundaries* of this electrophoresis image are computed (Figure 14.1, top-right). Notice that all closed curves found are indeed perceptually relevant since they surround the conspicuous spots. Many other possible boundaries in the noisy background have been ruled out and remain "masked in texture". Let us apply a second layer of grouping laws. This second layer will use as atomic objects the blobs found at the first step. For each of the detected boundaries three qualities will be computed:

1. *The area* enclosed by the boundary, whose histogram is displayed on the bottom left of Figure 14.1. There is a unique maximal mode in this figure, which actually groups all the blobs with similar areas and rules out two tiny blobs and a larger one enclosing two different blobs. Thus, almost all blobs get grouped by this quality, with the exception of two tiny spots and a double spot.
2. *The orientation* of each blob, an angle between $-\pi/2$ and $\pi/2$. This histogram (Figure 14.1, bottom-middle) again shows a single maximal mode, again computed by the formula of the third row of Table 14.1. This mode appears at both endpoints of the interval, since the dominant direction is $\pm\pi/2$ and these values are identified modulo π. Thus, about the same blobs as in quality 1 get grouped by their common orientation.
3. *The average gray level* inside each blob: Its histogram is shown on the bottom right of Figure 14.1. Again, most blobs, but not all get grouped with respect to this quality.

A further structural grouping law can be applied to build subgroups of blobs formed by alignment. This is illustrated in Figure 14.2 (bottom-right), where the meaningful alignments are found. This experiment illustrates the usual strong collaboration of partial gestalts: Most salient objects or groups come into sight by several grouping laws.

One would like to claim that the gestalt analysis of this apparently simple digital image is complete. This is not quite the case. A visual exploration shows that the fainter blobs, visible though they are, were not detected. They are not surrounded by a meaningful boundary. They are seen because of their similarity in shape and position to the detected gestalts. Indeed, these blobs are no longer visible when they are displaced in a more arbitrary position in the image.

14.3 The Limits of Every Partial Gestalt Detector

14.3.1 Conflicts Between Gestalt Detectors

This book argues in favor of a very simple principle, the Helmholtz principle, applicable to the automatic and unsupervised detection of any partial gestalt. Are the obtained detections in agreement with our perception? In this subsection, we will see several experiments proving that there is a good deal of visual illusion in any positive result provided by a partial gestalt detector. Partial gestalts often collaborate.

Fig. 14.3 Left: the INRIA pattern (size 512×512, from INRIA-Robotvis database). Middle: meaningful boundaries. Right: meaningful alignments. This experiment illustrates the possible collaboration of partial gestalts. The rectangles are detected as constrasted boundaries alone, and they also appear as a consequence of the alignment detector.

Fig. 14.4 Smooth convex sets or alignments?

In other words, a group detected by a partial gestalt is corroborated by another. For instance, the boundaries and alignments in Figure 14.3 are in good agreement. However, what can be said about the experiment in Figure 14.4? In this portrait of a cheetah, the alignment detector was applied. It worked well on straight grass leaves. Unfortunately, some unexpected alignments appear in the cheetah's fur. Is that a wrong detection? These alignments actually do exist. Each detected line is tangent to several convex dark spots on the fur. This tangency generates a meaningful excess of aligned points on the line, the convex sets being smooth enough and therefore having on their boundary a long enough segment tangent to the detected line.

An illustration of this fact is given in Figure 14.5. Four circles create two alignments on the two straight lines tangent to the circles. To geometers these alignments make sense but not to common perception. In fact such alignments are masked by the more powerful partial gestalts in game here – those that let us see circles. The right partial gestalts to deal with the circles are convexity, closure (of a curve), and

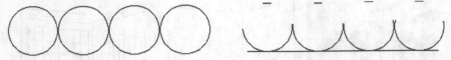

Fig. 14.5 Alignment is masked by good continuation and convexity: The small segments on the right are perfectly aligned. Any alignment detector should find them. All the same, this alignment disappears on the left figure, as we include the segments into circles. In the same way, the casual alignments in the Cheetah fur (Figure 14.4) are caused by the presence of many oval shapes. Such alignments are perceptually masked and should be computationally masked!

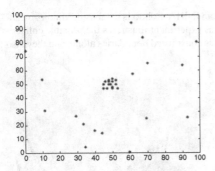
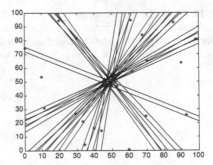

Fig. 14.6 One cluster or several alignments?

good continuation – or more prosaically, a circle detector. To discard the detected alignments, these other partial gestalts should be systematically searched when we look for alignments. Then the alignments that occur only because they are tangent to one or several smooth curves could be inhibited by the exclusion principle.

An alignment can be retained only if it does *not* derive from the presence of several smooth curves. This statement can be generalized. Indeed, no gestalt is just a positive quality. The outcome of a partial gestalt detector is valid only when *all* other partial gestalts have been tested and the eventual conflicts handled.

The same argument applies to the Figure 14.3.1 experiment. In that case, a dense cluster of points is present. Thus, it creates a meaningful amount of dots in many strips and the result is the detection of obviously wrong alignments. Again, the detection of a cluster should inhibit such alignment detections. We defined an alignment as "many points on a thin strip", but must add to this definition: "provided these points do not build one or two dense clusters".

14.3.2 Several Straight Lines or Several Circular Arcs?

One can reiterate the same problematic with another gestalt conflict. In Figure 14.7, a circle arc detector has been applied. This detector can be built on exactly the same principles as an alignment detector. First, compute the total number of arcs of circles in an image with given size $N \times N$. This number is roughly N^5, as the

Fig. 14.7 Left: original "MegaWave" image. Right: a circular arc detector is applied to the image. This image contains many smooth curves and obtuse angles. One can find meaningful circular arcs tangent to these structures. This illustrates the need for principles governing the interaction of partial gestalts. The best explanation for the observed structures is "good continuation" in the gestaltic sense (i.e., the presence of a smooth curve, or of straight lines (alignments) forming obtuse angles). Their presence entails the detection of arcs of circles, which are not the final explanation.

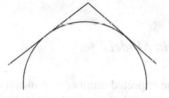 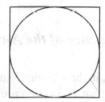

Fig. 14.8 Left: every obtuse angle can be made to have many points in common with some long arc of circle. Thus, an arc of circle detector will make wrong detections when obtuse angles are present (see Figure 14.7). In the same way, a circle detector will detect the circle inscribed in any square and, conversely, a square detector will detect squares circumscribed to any circle.

product of N^2 for the choice of the center of the circle, N for its radius, N for the starting point of the arc, and N for the ending point. At each point of the image, compute, as for alignments, an orientation. Let C be an arc of circle, and let l be its length (counted in independent points, at distance 2 from each other). Count the number k of points among the l whose orientation is aligned to a fixed precision p with the tangent to the circle. The event's NFA is $\text{NFA}(l,k,p) = N^5 \mathcal{B}(l,k,p)$. For the standard precision $p = 1/16$, all meaningful circle arcs of a digital image have been computed in Figure 14.7. Since the MegaWave image contains many smooth boundaries and several straight lines, many meaningful circular arcs are found. It may be discussed whether those circular arcs are present or not in the figure. Clearly, any smooth curve is locally tangent to some circle. In the same way, two segments with an obtuse angle are tangent to several circular arcs (see Figure 14.8). This can lead to a "wrong" detection of an arc of circle, where the detection of two alignments should be preferred. See, for example, the detected arc of circle on the coat, just under the "M". In fact a circle masks its tangent alignments. Conversely, we do not see an arc of circle where we see an obtuse angle. We can just hint here that an obvious solution to this conflict might be to choose as the best explanation the one with the lowest NFA, thus applying the *exclusion principle*. This implies the comparison of NFAs for different kinds of gestalt and raises the problem of a general calibration of NFA over many gestalts. For the time being, the statement of principles leading to such a cross-calibration is open.

In any case, *we cannot hope for any reliable explanation of a figure by summing up the results of one or several partial gestalts. Only a global synthesis, treating all conflicts of partial gestalts, can give the correct result.*

In view of these experimental counterexamples, it may well be asked why partial gestalt detectors often work so well. This is due to the redundancy of gestalt qualities in natural images. Indeed, most natural or synthetic objects are simultaneously contrasted with respect to background and with smooth boundary. They have straight or convex parts, and so forth. Thus, in many cases, each partial gestalt detector will lead to the same group definition. The experiments on the electrophoresis image (Figure 14.1) have illustrated this *collaboration of gestalt* phenomenon. In that experiment, partial gestalts collaborate and seem to be redundant.[1] This is an illusion that can be broken when partial gestalts do not collaborate.

14.3.3 Influence of the A-contrario Model

By definition, in the a-contrario model, the expected number of ε-meaningful structures is less than ε. In the case of alignments, the a-contrario model is a pure Gaussian noise image in which orientations are independent (at distance larger than 2) and uniformly distributed on $[0, 2\pi)$. In Chapter 5, Figure 5.2, we have shown such an example of a pure noise image. In this image, no meaningful alignement was detected. Now, if we quantize this image, on five gray levels for instance, many meaningful alignements are detected (see Figure 14.9). These alignements are mainly diagonal. The reason for this is that in a quantized noise image, the histogram of orientations is not flat anymore. The values multiple of $\pi/4$ are highly favored (the extreme case is the one of a binary noise image where the only possible orientations, with the computation of the gradient we have given in Chapter 5, are the multiples of $\pi/4$). The alignements detected in Figure 14.9 cannot be considered as false detections; they are part of the structure of this quantized image. Now, if we change the a-contrario model and replace the uniform assumption for orientations by the histogram of orientations in a quantized noise image, then no meaningful alignment will be detected.

When changing the a-contrario model, what we change is the number of false alarms (NFA) of the event we consider. In the example of Figure 14.9, all of the

---

[1] Recently, Krüger and Wörgötter [KW02] gave strong statistical computational evidence in favor of a collaboration between partial gestalt laws, namely collinearity, parallelism, color, contrast, and similar motion.

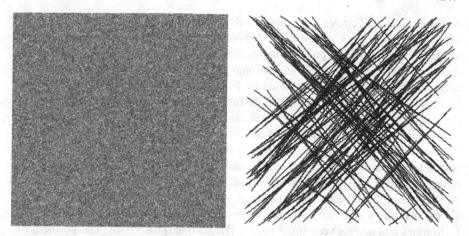

Fig. 14.9 Left: a pure Gaussian noise image quantized on five gray levels. Right: all meaningful alignements detected in this image.

detected alignements have a NFA larger than 10^{-5} under the uniform assumption. However, when using the histogram of orientations in the quantized image as an a-contrario assumption, all of the segments have a NFA larger than 1.

This is a general behavior of the a-contrario approach: If a particular structure is taken-into account in the a-contrario model, then this structure will not be meaningful anymore. This is what happens for alignments in a quantized image: If the a-contrario assumption includes the fact that diagonal orientations have a high probability, then diagonal alignments in the quantized noise image are not meaningful anymore.

As pointed out to us by David Mumford, the same kind of phenomenon arose in the detection of repeating spike patterns in neural recordings (see, for instance, [CRM03] or [HGAB03]). It turned out that apparently very significant repetitions of complex spike patterns were caused by the null hypothesis (the a-contrario model) not allowing bursts. If bursts were included, these repetitions were not significant at all. This is also illustrated by the example discussed in Figure 14.3.1: If the a-contrario model used for the detection of dots alignments includes the fact that dots can be organized into clusters, then no meaningful alignment will be detected in this set of dots.

14.4 Bibliographic Notes

This chapter is directly inspired from the papers [DMM03b, DMM01a, DMM04]. In Christopher Small's book [Sma96] Section 6.3, the author provides an analysis of post mold data. The problem is to derive from these data the shape of a village of circular huts. The problem is treated in exactly the way we treated the detection of arcs of circles by computing the probability that just by chance a number larger than k posts among the l observed in the data falls into a given crown. Since a first draft of the present manuscript was completed, several studies were done to improve the list of computational gestalts. The Chapter 9 boundary and edge detection theory was completed by F. Cao, P. Musé, and F. Sur in [CMS05]. In [Cao04] and [Cao03] F. Cao successfully applied the Helmholtz principle for the detection of good continuations and corners in image level lines. In [MVM05], L. Moisan, E. Villéger, and J.M. Morel worked on the *constant width* detection. F. Cao, Y. Gousseau, P. Musé, and F. Sur developed a theory of shape recognition with a learned a-contrario model which makes it a very flexible tool for shape databases (see [MSCG03], [MSC+06a], [MSC+06b], [CGM+04], [Mus04], [Sur04]). A. Desolneux, A. B. Petro, and J. Delon generalized the Chapter 7 histogram analysis method into a nonparametric law estimation theory. This theory segments 1-D histograms and detects very small modes [DDLP07b, DDLP07a]. The problem of meaningful clusters has been addressed in higher dimension by F. Cao, J. Delon, A. Desolneux, P. Musé, and F. Sur in [CDD+04]. They define maximal meaningful clusters in a hierarchical clustering tree and apply the resulting algorithm to shape recognition. The Helmholtz principle was used (in a way very similar to the alignment case of Chapter 5) by J.L. Lisani and J.M. Morel to detect major changes in satellite images [LM03] and by F. Cao and P. Bouthemy in [CB05] for image similarity detection. G. Koepfler, F. Dibos, and S. Pelletier used similar techniques for real-time segmentation of moving objects in a video sequence [KDP05]. The problems of motion detection and estimation using a-contrario models were also addressed by L. Igual, L. Garrido, and V. Caselles in [IGC05] and by T. Veit, F. Cao, and P. Bouthemy in [VCB05] and [VCB04]. The work of A. Robin, L. Moisan, and S. Le Hégarat-Mascle [RMHM05] treats the problem of automatic land-cover change detection from coarse resolution images by the a-contrario approach.

Chapter 15
Other Theories, Discussion

In this chapter, we review and discuss precursory and alternative theories. We start in the first section with Lindenbaum et al. Their papers contain a theory of shape detection whose setting is essentially the same as the one developed in this book. The Bienenstock et al. compositional model discussed in Sections 15.2 and 15.5 is an ambitious theory attempting to build directly a grammar of visual primitives. A nice illustration of these compositional approaches is the work of Zhu et al. described in Section 15.2. Section 15.3 discusses the link among meaningful events, hypothesis testing, and Signal Detection Theory. It also shows that the Number of False Alarms (NFA) can be put in a classical statistical framework where multiple testing is involved. In Section 15.4 the Arias-Castro et al. geometric detection theory is addressed. This theory is very close in spirit to the tools in this book and are actually partly inspired from it. It gives complementary information on asymptotic geometric detection thresholds and hints on how to speed up detection algorithm. Section 15.5 discusses the Bayesian theory according to which the probability of the image interpretation given the observation must be maximized. An extension of this theory, the Minimum Description Length, is also invoked in the compositional model. In both cases, a probability is maximized. In contrast, meaningful events were obtained by minimizing an a-contrario probability. This point is discussed and the complementarity of both approaches are indicated.

15.1 Lindenbaum's Theory

We already mentioned D. Lowe and J. Stewart, who proposed methods very close in spirit to the detection theory developed in this book. Other precursors are Michael Lindenbaum and his collaborators [Lin97]. They proposed evaluating the absolute performance of an invariant shape recognition device. This performance depends on: the number of points k in the shape, the number of points N in the background, the accuracy d of the recognition, the required invariance, and ε, the allowed error probability for each test. The typical theorems proved by Lindenbaum involve a Hoeffding inequality to give a lower bound for the minimal number of points k in the shape necessary to get an $1 - \varepsilon$ confidence for the recognition. Of course, k is a function of the four mentioned variables. Fixing the accuracy d leads to evaluate what we called the *number of tests*. Of course, this number of tests grows with the invariance requirement and gets larger when the accuracy decreases. Lindenbaum points out the *log* dependence of the threshold k on N, d, and δ.

15.2 Compositional Model and Image Parsing

In [BGP97] E. Bienenstock, S. Geman, and D. Potter proposed a gestalt *compositional model* in which visual primitives are recursively composed subject to syntactic restrictions, to form tree-structured objects and object groupings. The ambiguities of the final result are solved by a minimum-description-length functional. For these authors, the compositional rules have a structure close to Chomsky's grammar. In the probabilistic modeling of these rules, the authors rely on Laplace's Essay on Probability.

> On a table we see letters arranged in this order, Constantinople, *and we judge that this arrangement is not the result of chance, not because it is less possible than the others, for if this word were not employed in any language we should not suspect it came from any particular cause, but this word being in use amongst us, it is incomparably more probable that some person has thus arranged the aforesaid letters than this arrangement is due to chance.*

For Bienenstock, Geman, and Potter, the fact that Constantinople is in use in French and other languages constitutes a binding rule

$$C + o + n + s + t + a + n + t + i + n + o + p + l + e \Rightarrow Constantinople$$

and this binding is by far more likely than the one building $Ipctneolnosant$, which has the same letters. In our terms, Laplace is implicitly using an a-contrario model, where any combination of the 26 alphabet letters would be equally likely. A modern dictionary contains not more than 10^5 words. The number of possible words with 14 letters like Constantinople is about 2.4×10^{19}. In this background model, the probability of the group Constantinople happening "just by chance" is less than 10^{-14}. We could have called the Helmholtz principle the Laplace principle! The Bienenstock et al. theory is close to the theory presented in this book because the binding rules are learned as "suspicious coincidences". Suspicious coincidences are meaningful events in an a-contrario model. The scope of the theory is more ambitious. The theory aims at a global image explanation and at an automatic definition of partial gestalts. The fact that alignments are of interest in an image should be discovered and not imposed *a priori* as we did in Chapter 5.

A nice illustration of these ideas of compositional models and grammars is the work of Han and Zhu [HZ05] for image parsing on scenes with man-made objects, such as buildings, hallways, kitchens, and livingrooms. Their model includes more gestalt grouping principles than any other to date. Its combines together alignements, vanishing points, parallelism, perspective, with recursivity. The objective of image parsing is to decompose an image into its constituent components in a hierarchical structure represented by a parsing graph. This is a natural idea, and one can ask why this problem has not been fully addressed before. There are different reasons for this: (1) It is not always easy to compute "primitive patterns" (like segments, or rectangles, or edges) in a reliable way from images. (2) Many conflicts are possible when building up higher-order objects from the primitives. Thus, one needs a mathematical model that is able to handle together and compare different conflicting interpretations. (3) The problem of finding the best interpretation of a scene is the same as finding the best parsing graph, and powerful inference algorithms are needed to find the solution. Recent developments about attribute graph grammars and stochastic context-sensitive grammars, and also inference tools like data-driven Monte Carlo Markov chains (MCMC), have made it possible to apply grammar parsing algorithms to real images and have motivated the work of Han and Zhu. Let us describe their work with more details and explain how they combine elementary gestalt grouping principles for parsing "rectangular scenes". Their algorithm proceeds in different phases. Some phases are bottom-up: From small straight segments and vanishing points detected in the input image, some "strong" rectangles are proposed. And some phases are top-down: Grammar rules are used, and rectangles produce top-down proposals that will be, or not, validated from the edge map. See Figure 15.1 for an illustration of this. There are four attribute grammar rules for nonterminal nodes, and they can be used recursively: There is the line production rule, the cube production rule, the mesh production rule, and the nesting production rule. Each production rule is associated with equations that constrain the attributes (like position, size, and orientation) of the node and of its children. Horizontal connections between nodes also show constraints between attributes. The problem of finding the best parsing graph G of an image I is expressed in a Bayesian framework, that is, find

$$G^* = \arg\max p(I|G)p(G).$$

The prior $p(G)$ is computed as a function of the gain in bits when coding the attributes of children nodes given the attributes of the parent node. The likelihood $p(I|G)$ is computed thanks to the primal sketch model of Guo, Zhu and Wu [GZW03]. To infer the best parsing graph, MCMC methods can be used. In the case of rectangular scenes, since the parsing graph is not too big (typically 2–3 layers and less than 20 nodes), more direct search algorithms can be used. An example of the type of results obtain by Han and Zhu is shown in Figure 15.2.

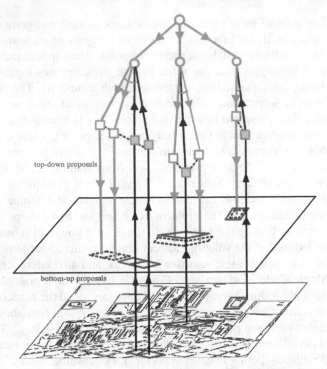

top-down proposals

bottom-up proposals

Fig. 15.1 Work of Han and Zhu about parsing "rectangular scenes": Some phases of their algorithm are bottom-up ("strong" rectangles are proposed from the edge map), and some phases are top-down (rectangles are proposed by some grammar production rules). (Figure courtesy of S.C. Zhu.)

Fig. 15.2 Example of results of Han and Zhu. From left to right: input image; edge map; rectangles detected and grouped. (Figures courtesy of S.C. Zhu.)

15.3 Statistical Framework

15.3.1 Hypothesis Testing

The notion of ε-meaningful has to be related to the classical statistical framework of hypothesis testing. The a-contrario model can be seen as the null hypothesis H_0 and the significance level of the test is $\alpha = \varepsilon/N_{tests}$, where N_{tests} denotes the number of geometric configuration that are tested (we have, for instance, $N_{tests} = N^4$ in the case of segments in an image of size $N \times N$ pixels). The null hypothesis is rejected when the p-value of the test is less than the significance level α. In the case of alignments, the a-contrario model or null hypothesis is that the orientations along a segment are independent and uniformly distributed on $[0, 2\pi)$. Segments that reject the null hypothesis are called ε-meaningful segments, or "detections". Now, the difference that leads us to have a slightly different terminology is the following: We are not in a position to assume that the segment detected as ε-meaningful are independent in anyway. Indeed, if (for example) a segment is meaningful, it may be contained in many larger segments, which also are ε-meaningful. Thus, it is convenient to compare the number of detected segments to the expectation of this number. This is not exactly the same situation as in failure detection, where the failures are somehow disjoint events. This means that ε is an absolute parameter, not depending on the size of the image but only on the number of false detections that the user allows. Of course, if the image is larger, it may be expected that an increasing number of false detections should be allowed. However, by fixing ε always smaller than 1, we decided not to take this opportunity. Our proposed definition of meaningfulness is also related to the statistical analysis of functional medical images (fMRI, PET) by Statistical Parameter Maps (SPM), with two main differences, however. The first one is this: In the recent work of Stuart Clare (FMRIB Center, Oxford, see [Cla97]) and in the works of Friston et al. [FFLF91] and Forman et al. [FCF+95], an hypothesis testing method against white noise is performed in a time series. As in the present work, the binomial law appears and a careful account of the effect of filtering on the number of effective degrees of freedom: This leads, for example, Clare to divide this number by 3 after a small Gaussian filtering and is related to our decision of considering only nets of points at a distance larger than 2. Clare does as we do; he p-tests against the white noise assumption and admits a p-value of 0.005 by patient. Here is the main difference: The number of patients and the length of the data are not taken into account in the test. In particular, the time length of the test is, of course, just enough to perform a significant test and the p-value is a threshold "per patient". In our case, we have two factors: The first one is that the number of "patients" is huge. Thus, with a p-test, the expectation of false detections would be much above 1, which is what we avoid by imposing ε much smaller than 1 and by entering into the computation the number of segments N^4. This is why we compute an expectation and not a probability: We have too many and not independent trials. The reason for introducing expectation here is the nonindependence (contrarily to patients) and the huge number of trials, increasing with the image size.

15.3.2 Various False Alarms or Error Rates Compared to NFA

The computation of a number of false alarms is related to the classical statistical question of multiple testing (see, for example, [HT87]). Let N_{tests} be the number of tests and let V be the random variable that denotes the number of false alarms (also called false positives). The terminology distinguishes the following:

- FWER: Familywise error rate, probability of at least one false alarm $= \mathbb{P}(V \geq 1)$.
- PFER: Per family error rate, expectation of the number of false alarms $= \mathbb{E}[V]$.
- PCER: Per comparison error rate, expectation of the proportion of false alarms among the total number of tests $= \mathbb{E}[V/N_{tests}]$.

From the statistical viewpoint, the NFA can be viewed as a decision threshold taking into account the multiplicity of tests. It is related to the *Per Family Error Rate* (PFER). The consistency of the NFA definition relies on the result *the expected number of false alarms of ε-meaningful events is less than ε*. Various versions of this simple result have been given. We can view the question "Is the NFA of some event less than ε?" as a test. Considering this family of tests, we can also consider its PFER, namely the expectation of the number of positive answers. The above-mentioned generic result can be denoted as

$$\text{PFER (NFA} \leq \varepsilon) \leq \varepsilon. \tag{15.1}$$

This property can be used as a starting point to define *NFA*s, as done in [GM06], and for continuous random variables, (15.1) is generally an equality (see Section 5.6).

Let us formalize a bit more. Call $\mathcal{E} = (E_1, \ldots, E_{N_{tests}})$ the Boolean random values equal to 1 if the NFA of the observed configuration i is less than ε (which is equivalent to say that the p-value of the observed configuration is less than ε/N_{tests}). Thus, $E_i = 1$ if and only if $\text{NFA}(i) = N\mathbb{P}(E_i = 1) \leq \varepsilon$. The FWER is defined as the probability that *at least one* of the tests is positive. Thus,

$$\text{FWER}(\mathcal{E}) = \mathbb{P}(\max_i E_i \geq 1).$$

By the Bonferroni inequality,

$$\text{FWER}(\mathcal{E}) \leq \sum_{i=1}^{N_{tests}} \mathbb{P}(E_i = 1) = \sum_{i=1}^{N_{tests}} \frac{\text{NFA}(i)}{N_{tests}} \leq \sum_{i=1}^{N_{tests}} \frac{\varepsilon}{N_{tests}} \leq \varepsilon.$$

All in all:

Theorem 10 *Consider a family of N_{tests} configurations indexed by i for which a NFA is computed. Consider the family $\mathcal{E}$ of the following N tests: "is configuration i ε-meaningful?" Then*

$$\text{PCER}(\mathcal{E}) = \frac{\varepsilon}{N_{tests}}, \quad \text{FWER}(\mathcal{E}) \leq \varepsilon, \quad \text{PFER}(\mathcal{E}) \leq \varepsilon. \tag{15.2}$$

We thank Jérémie Jakubowicz for valuable suggestions on the comparison of these various error rates.

15.3.3 Comparison with Signal Detection Theory

Compared to the classical Signal Detection Theory [Tre68, Kay98], we believe that the approach we proposed in this book may be an interesting alternative, especially when human vision is concerned. This point has been recently discussed in [GM06] in the context of medical imaging analysis, for which metrics reproducing the performance of human vision are often sought.

In Signal Detection Theory, both the background and the signal are modeled with probability models, called H_0 and H_1, respectively. Then an Ideal Observer is defined from the likelihood ratio of the observed data under H_1 and H_0, and the Neyman-Pearson Lemma ensures that this Ideal Observer cannot be outperformed. Several mathematical difficulties may arise if H_0 and H_1 are not simple enough, but approximations of the Ideal Observer (like the Prewhitening Matching filter, for example) can be used to overcome this difficulties. A more fundamental issue is: How to use Signal Detection Theory when the signal to detect is not known precisely, as it is often the case in Computer Vision? The a-contrario approach proposes an answer, which consists in breaking the symmetry between the background and signal models. The structure to be detected is no longer described by a probability measure H_1, but by one or several measurements inspired from Gestalt Theory. In the case of alignments, for example, the measurement is simply the number of aligned directions on a segment, which is clearly a significant cause of perceptual grouping. Describing grouping laws that way is easier and more flexible than trying to build a probabilistic model of all potential structures, as required in the H_1 formulation. This is an agreement with general principles of vision (human or not), which do not build a dictionary of the visual world but, rather, a set of rules to understand it. In fact, there is no theoretical contradiction between Signal Detection Theory and the a contrario approach, since the latter may use the Ideal Observer to derive the optimal measurements associated to a given (supposedly known) H_1 model of the signal. However, this is not appropriate in general, as shown in [GM06] for example: When "spots" are to be detected on a low-frequency texture, the Ideal Observer builds correlation templates that are not L^2 images but distributions, which are very sensitive to modeling errors. These matching templates perform optimally if the searched signal corresponds well to the model, but poorly if the signal model is wrong. Conversely, a-contrario observers based on simpler measurements are more robust and often nearly optimal.

15.4 Asymptotic Thresholds

A Stanford statistical geometry group consisting of E. Arias-Castro, D. Donoho, X. Huo and C. Tovey has developed a geometric detection statistical theory close in spirit and sometimes directly inspired by the theory presented in this volume. Their aim is to find the minimal number of geometric tests for each detection. In the case of alignments, this theory yields a reduction from N^4 tested segments to only $N^2 \log N$. Unfortunately, the model is a random dot Poisson model, not directly applicable to digital images as the model of Chapter 5. The authors prove the existence of asymptotic geometric detection thresholds when $N \to \infty$ for the *good continuation* gestalt. Here are some appealing examples. Assume n points are scattered uniformly on the unit square $[0, 1]^2$ by a Poisson spatial process. Then the following hold:

- The asymptotic maximal number of points on a length l curve is $O(l\sqrt{n})$.
- The asymptotic maximal number of points in a twice differentiable curve with *a priori* bounded length and curvature is $O(n^{1/3})$.
- Take n random points endowed with a random uniform direction, then the asymptotic maximal number of points in the same type of curve is $O(ln^{14})$ [ACDHT05].
- There is an asymptotic optimal detection threshold for a line segment with different densities in a Gaussian white noise image. Here, the threshold is performed on the density of the segment [ACDH05].
- There are asymptotic optimal detection thresholds for a different density on a $C^{\alpha,\beta}$ curve in Poisson noise. Here, $1 < \alpha < 2$ is a Hölder exponent [ACDH03].

These papers consider asymptotic thresholds as the number of samples tends to infinity. Indeed, one of the main ingredients is the Erdös-Renyi asymptotic estimate on the size of the longest run of heads in a series of m Bernoulli trials. For practical applications, the nonasymptotic detection thresholds can simply be learned on white or Poisson noise images.

15.5 Should Probability Be Maximized or Minimized?

The present essay on computational Gestalt Theory has tried to build a mathematical framework for the idea that significant geometric structures in an image correspond to low-probability events in noise, which can be detected by parameterless algorithms (see Figure 15.3). Maximal meaningful events were defined as event with the smallest probability in a class of meaningful events.

Most gestalt detection theories actually involve a probabilistic functional that is to be *maximized, not minimized!* These theories of Bayesian inspiration assume that the images contain geometric structures and try to estimate their best location and description. Let us take as a generic example the segmentation problem. The general idea of segmentation algorithms is to minimize a functional of the kind

$$F(u, u_0) + R(u),$$

Fig. 15.3 The church of Valbonne: size 502×480 (from INRIA-Robotvis database). The detection of alignments and boundaries is made in both cases with $\varepsilon = 1$ and supports the idea that parameterless methods are possible for image analysis, in contrast with variational methods.

where u_0 is the given image defined on a domain $\Omega \subset \mathbb{R}^2$, $F(u, u_0)$ is a fidelity term, and $R(u)$ is a regularity term. F and R define an *a priori* model. Here are two formalisms for it: one deterministic and the other one probabilistic:

- In the Mumford-Shah model (see [MS85], [MS89], [NMS93]) the energy functional to be minimized is

$$E(u, K) = \lambda^2 \int_{\Omega \backslash K} |\nabla u|^2 \, dx + \mu \lambda^2 \text{length}(K) + \int_{\Omega - K} (u - u_0)^2 \, dx,$$

where u is the estimated image, K is its discontinuity set, and the result (u, K) is called a segmentation of u_0 (i.e., a piecewise smooth function u with a set of contours K).
- In the Bayesian model (see [GG84]), $\mathbf{y} = (y_s)_{s \in S}$ denotes the observation (the degraded image). The aim is to estimate the real image $\mathbf{x} = (x_s)_{s \in S}$ knowing that the degradation model is given by a conditional probability $\Pi(\mathbf{y}|\mathbf{x})$ and that the *a priori* law of $\mathbf{x}$ is given by a Gibbs distribution $\Pi(\mathbf{x}) = Z^{-1} \exp(-U(\mathbf{x}))$ (for binary images, the main example is the Ising model). We then have to find the MAP (Maximum a Posteriori) of

$$\Pi(\mathbf{x}|\mathbf{y}) = \frac{\Pi(\mathbf{y}|\mathbf{x})\Pi(\mathbf{x})}{\Pi(\mathbf{y})}.$$

Assume that $\Pi(\mathbf{y}|\mathbf{x}) = C \exp(-V(\mathbf{x}, \mathbf{y}))$. For example, in the case of a Gaussian noise

$$\Pi(\mathbf{y}|\mathbf{x}) = \left(\frac{1}{2\pi\sigma^2}\right)^{|S|/2} \exp\left(-\frac{1}{2\sigma^2} \sum_{s \in S} (y_s - x_s)^2\right),$$

finding the MAP is equivalent to seeking for the minimum of the functional

$$V(\mathbf{x}, \mathbf{y}) + U(\mathbf{x}).$$

The Bayesian methods first build up a probability model of a pattern of interest. Then the likeliest event for the model in a given image is sought. Probably the best exponent of this method is Ulf Grenander [Gre93]. We quote his programme extracted from the Brown University pattern theory group website (HTTP://WWW.DAM.BROWN.EDU/PTG/):

The Brown University pattern theory group is working with the belief that the world is complex, and to understand it, or a part of it, requires realistic representations of knowledge about it. We create such representations using a mathematical formalism, pattern theory, that is compositional in that the representations are built from simple primitives, combined into (often) complicated structures according to rules that can be deterministic or random. This is similar to the formation of molecules from atoms connected by various forms of bonds. Pattern theory is variational in that it describes the variability of the phenomena observed in different applications in terms of probability measures that are used with a Bayesian interpretation. This leads to inferences that will be realized by computer algorithms.

This program is gestaltist in that it proposes to build pattern models from simple primitives by a compositional representation. These pattern models associate with each observed image or configuration in an image a probability that is thereafter maximized. The second part of the program explicitly suggests a variational framework.

In practice, one is faced with two main problems when working with variational methods. First, even in the simple case of image segmentation, there are normalization constants (λ, μ, etc.) and the result depends very much on the value of these constants. Notice that in the second-mentioned Bayesian model, U contains several parameters and the resulting functional also depends on a degradation model. Second, variational methods always deliver a minimum for their functional. They do not yield any criterion to decide whether an obtained segmentation is relevant or not or whether a sought for pattern is present or not.

The probabilistic framework leading to the variational methods should, in principle, give a way to estimate the parameters of the segmentation functional. In the deterministic framework, these parameters can sometimes be estimated as Lagrange multipliers when (for example) a noise model is at hand, as in the Rudin-Osher-Fatemi method (see [ROF92]). It is nonetheless true that variational methods necessarily propose a rough and incomplete model for real-world images. Their parameters are generally not correctly estimated. This leads to supervised methods.

Another possibility, which turns out to be a significant improvement of Bayesian methods, is the Minimum Description Length (MDL) method introduced by Rissanen [Ris83] and first applied in image segmentation by Yvon Leclerc [Lec89]. This last method, applied to detect regions and their boundaries in an image, lets us automatically fix the weight parameters whose presence we criticized in the Mumford-Shah model. The resulting segmentation model remains, all the same, unproved: The MDL principle does not prove the existence of regions. It only gives their best description, provided the image indeed can be segmented into constancy regions. This fact is easily explained. The MDL principle assumes that a model, or a class of models, is given and then computes the best choice of the model parameters and of the model explaining the image. As far as perception theory is concerned, more is needed, namely a proof that the model is the right one.

Not all geometric detection methods are variational. Other classical and complementary examples are the Hough Transform (see [Mai85]), the detection of globally salient structures by Sha'Ashua and Ullman (see [SU88]), the Extension Field of Guy and Medioni (see [GM96]), and the Parent and Zucker curve detector (see [PZ89]). These methods have the same drawback as the variational models of segmentation described earlier. The main point is that they *a priori* suppose that what they want to find (lines, circles, curves, etc.) is in the image. They may find too many or too little such structures in the image and do not yield an **existence proof** for the found structures.

These comments do not rule out variational methods. We have indeed seen that a comparison of NFAs might lead to the solution of gestalt conflicts by always choosing the detected gestalt with lowest NFA. The theory and experiments of this book emphasize the fact that the detection of structure has an intermediate stage, clearly missed in the variational framework: *Before we look for the most likely structures, we have to make a list of all proven structures.* So we will close the discussion by proposing a slightly different role to variational methods. We have shown in this book that partial gestalts can be computed by the Helmholtz principle followed by a maximality argument and/or an exclusion principle. The discussions of gestaltists about "conflicts of gestalts", so vividly explained in the books of Kanizsa, might well be solved by a few information-theoretical principles. As we mentioned earlier, their solution will lead us back to a variational framework, as was widely anticipated by gestaltists themselves.

References

[AADLTT06] I. Abraham, R. Abraham, A. Desolneux, and S. Li-Thiao-Té. Significant edges in the case of non-stationary gaussian noise. Technical report, MAP5, Paris Descartes University, 2006. To appear in *Pattern Recognition* 2007.

[ABE$^+$55] M. Ayer, H.D. Brunk, G.M. Ewing, W.T. Reid, and E. Silverman. An empirical distribution function for sampling with incomplete information. *Annals of Mathematical Statistics*, 26(4):641–647, 1955.

[Abu89] A.S. Abutaled. Automatic thresholding of gray-level pictures using two-dimensional entropy. *Computer Vision, Graphics and Image Processing*, 47:22–32, 1989.

[ACDH03] E. Arias-Castro, D.L. Donoho, and X. Huo. Adaptive multiscale detection of filamentary structures embedded in a background of uniform random points. Technical report, Department of Statistics, Stanford University, CA, 2003.

[ACDH05] E. Arias-Castro, D.L. Donoho, and X. Huo. Near optimal detection of geometric objects by fast multiscale methods. *IEEE Transactions on Information Theory*, 51(7):2402–2425, 2005.

[ACDHT05] E. Arias-Castro, D.L. Donoho, X. Huo, and C. Tovey. Connect-the-dots: How many random points can a regular curve pass through? *Advances in Applied Probability*, 37:571–603, 2005.

[ADV03] A. Almansa, A. Desolneux, and S. Vamech. Vanishing points detection without any a priori information. *IEEE Transactions on Pattern Analysis and Machine Intelligence*, 25(4):502–507, 2003.

[AFI$^+$06] A. Almansa, G. Facciolo, L. Igual, A. Pardo, and J. Preciozzi. Small baseline stereo for urban digital elevation models using variational and region-merging techniques. Technical report, Facultad de ingenieria, Montevideo, Uruguay, April 2006.

[Ale93] K. S. Alexander. Finite clusters in high-density continuous percolation: Compression and sphericality. *Probability Theory and Related Fields*, 97(1-2):35–63, 1993.

[Ana65] J. Anastassiadis. *Définition des fonctions eulériennes par des équations fonctionnelles*. Gauthier-Villars, Paris, 1965.

[AT00] M.E. Antone and S. Teller. Automatic recovery of relative camera rotations for urban scenes. In *International Conference on Computer Vision and Pattern Recognition*, volume II, pages 282–289, 2000.

[Att54] F. Attneave. Some informational aspects of visual perception. *Psychology Review*, 61:183–193, 1954.

[Bah60] R. Bahadur. Some approximations to the binomial distribution function. *Annals of Mathematical Statistics*, 31:43–54, 1960.

[Bar83] S.T. Barnard. Interpreting perspective images. *Artificial Intelligence*, 21:435–462, 1983.

[BBBB72] R.E. Barlow, D.J. Bartholomew, J.M. Bremner, and H.D. Brunk. *Statistical Inference Under Order Restrictions*. Wiley, New York, 1972.

[Ben62]   G. Bennet. Probability inequalities for the sum of independent random variables. *Journal of the American Statistical Association*, 57:33–45, 1962.

[BG05]   G. Blanchard and D. Geman. Sequential testing designs for pattern recognition. *Annals of Statistics*, 33(3):1155–1202, 2005.

[BGP97]   E. Bienenstock, S. Geman, and D. Potter. Compositionality, MDL priors, and object recognition. In *Advances in Neural Information Processing Systems*, volume 9, pages 838–844. M.C. Mozer, M.I. Jordan, and T. Petsche, eds, MIT Press, Cambridge, MA, 1997.

[Bir89]   L. Birgé. The Grenander estimator: A nonasymptotic approach. *Annals of Statistics*, 17(4):1532–1549, 1989.

[Bir97]   L. Birgé. Estimation of unimodal densities without smoothness assumptions. *Annals of Statistics*, 25(3):970–981, 1997.

[BJ83]   J.R. Bergen and B. Julesz. Textons, the fundamental elements of preattentive vision and perception of textures. *Bell System Technical Journal*, 62(6):1619–1645, 1983.

[BK53]   E. Brunswik and J. Kamiya. Ecological cue-validity of "proximity" and other gestalt factors. *American Journal of Psychology*, 66:20–32, 1953.

[BM95]   B. Boufama and R. Mohr. Epipole and fundamental matrix estimation using virtual parallax. In *International Conference on Computer Vision*, pages 1030–1036, 1995.

[BS96]   A.J. Bell and T.J. Sejnowski. Edges are the "independent components" of natural scenes. *Advances in Neural Information Processing Systems*, 9, 1996.

[BVZ01]   Y. Boykov, O. Veksler, and R. Zabih. Fast approximate energy minimization via graph cuts. *IEEE Transactions on Pattern Analysis and Machine Intelligence*, 23(11):1222–1239, 2001.

[Can86]   J.F. Canny. A computational approach to edge detection. *IEEE Transactions on Pattern Analysis and Machine Intelligence*, 8(6):679–698, 1986.

[Cao03]   F. Cao. Good continuation in digital images. In *International Conference on Computer Vision*, volume 1, pages 440–447, 2003.

[Cao04]   F. Cao. Application of the Gestalt principles to the detection of good continuations and corners in image level lines. *Computing and Visualisation in Science. Special Issue, Proceeding of the Algoritmy 2002 Conference*, 7:3–13, 2004.

[CB05]   F. Cao and P. Bouthemy. A general criterion for image similarity detection. Technical report, INRIA, 2005.

[CCJA94]   C.-I. Chang, K. Chen, J.Wang, and M. Althouse. A relative entropy-based approach to image thresholding. *Pattern Recognition*, 27(9):1275–1289, 1994.

[CCM96]   V. Caselles, B. Coll, and J.-M. Morel. A Kanizsa programme. *Progress in Nonlinear Differential Equations and Their Applications*, 25:35–55, 1996.

[CCM02]   V. Caselles, B. Coll, and J.M. Morel. Geometry and color in natural images. *Journal of Mathematical Imaging and Vision*, 16(2):89–105, 2002.

[CDD+04]   F. Cao, J. Delon, A. Desolneux, P. Musé, and F. Sur. An a contrario approach to clustering and validity assessment. Technical Report 2004-13, CMLA, ENS Cachan, 2004.

[CDD+07]   F. Cao, J. Delon, A. Desolneux, P. Musé, and F. Sur. A unified framework for detecting groups and application to shape recognition,. *Journal of Mathematical Imaging and Vision*, 27(2):91–119, 2007.

[CGM+04]   F. Cao, Y. Gousseau, P. Musé, F. Sur, and J.M. Morel. Accurate estimates of false alarm number in shape recognition. Technical Report 2004-01, CMLA, ENS Cachan, 2004.

[Che52]   H. Chernoff. A measure of asymptotic efficiency for tests of a hypothesis based on the sum of observations. *Annals of Mathematical Statistics*, 23:493–507, 1952.

[CKS97]   V. Caselles, R. Kimmel, and G. Sapiro. Geodesic active contours. *International Journal of Computer Vision*, 1(22):61–79, 1997.

[Cla97]   S. Clare. *Developing the technique of functional Magnetic Resonance Imaging to study visual, motor and auditory brain activation*. PhD thesis, University of Nottingham, England, October 1997.

[CMS05]   F. Cao, P. Musé, and F. Sur. Extracting meaningful curves from images. *Journal of Mathematical Imaging and Vision*, 22(2-3):159–181, 2005.

[CRM03]   Z. Chi, P.L. Rauske, and D. Margoliash. Detection of spike patterns using pattern filtering, with applications to sleep replay in birdsong. *Neurocomputing*, 52-54:19–24, 2003.

[CRZ00]   A. Criminisi, I. Reid, and A. Zisserman. Single view metrology. *International Journal of Computer Vision*, 40(2):123–148, 2000.

[CT91]   T.M. Cover and J.A. Thomas. *Elements of Information Theory*. Wiley, New York, 1991.

[Dav75]   L. Davis. A survey of edge detection techniques. *Computer Graphics and Image Processing*, 4:248–270, 1975.

[DDLP04]   J. Delon, A. Desolneux, J.L. Lisani, and A.B. Petro. Histogram analysis and its applications to fast camera stabilization. In *International Workshop on Systems, Signals and Image Processing*, 2004.

[DDLP07a]   J. Delon, A. Desolneux, J.L. Lisani, and A.B. Petro. Automatic color palette. *Inverse Problems and Imaging*, 1(2):265–287, 2007.

[DDLP07b]   J. Delon, A. Desolneux, J.L. Lisani, and A.B. Petro. A non parametric approach for histogram segmentation. *IEEE Transactions on Image Processing*, 16(1):253–261, 2007.

[Der87]   R. Deriche. Using Canny's criteria to derive a recursively implemented optimal edge detector. *International Journal of Computer Vision*, pages 167–187, 1987.

[dFM99]   J.-P. d'Alès, J. Froment, and J.-M. Morel. Reconstruction visuelle et généricité. *Intellectica*, 1(28):11–35, 1999.

[DH73]   R.O. Duda and P.E. Hart. *Pattern Classification and Scene Analysis*. Wiley, New York, 1973.

[DLMM02]   A. Desolneux, S. Ladjal, L. Moisan, and J.-M. Morel. Dequantizing image orientation. *IEEE Transactions on Image Processing*, 11(10):1129–1140, 2002.

[DMM00]   A. Desolneux, L. Moisan, and J. M. Morel. Meaningful alignments. *International Journal of Computer Vision*, 40(1):7–23, 2000.

[DMM01a]   A. Desolneux, L. Moisan, and J.-M. Morel. Automatic image analysis: a challenge for computer vision. In *Actes du GRETSI*, Toulouse, 2001.

[DMM01b]   A. Desolneux, L. Moisan, and J.-M. Morel. Edge detection by Helmholtz principle. *Journal of Mathematical Imaging and Vision*, 14(3):271–284, 2001.

[DMM03a]   A. Desolneux, L. Moisan, and J.-M. Morel. Computational Gestalts and perception thresholds. *Journal of Physiology*, 97(2-3):311–324, 2003.

[DMM03b]   A. Desolneux, L. Moisan, and J.-M. Morel. A grouping principle and four applications. *IEEE Transactions on Pattern Analysis and Machine Intelligence*, 25(4): 508–513, 2003.

[DMM03c]   A. Desolneux, L. Moisan, and J.-M. Morel. Maximal meaningful events and applications to image analysis. *Annals of Statistics*, 31(6):1822–1851, 2003.

[DMM03d]   A. Desolneux, L. Moisan, and J.-M. Morel. Variational snake theory. In S. Osher and N. Paragios, editors, *Geometric Level Set Methods in Imaging, Vision and Graphics*, pages 79–99. Springer-Verlag, New-York, 2003.

[DMM04]   A. Desolneux, L. Moisan, and J.-M. Morel. Gestalt theory and computer vision. In *Seeing, Thinking and Knowing*, pages 71–101. A. Carsetti ed., Kluwer Academic Publishers, 2004.

[Dos69]   F. Dostoievski. *Le joueur*. 1869.

[DZ93]   A. Dembo and O. Zeitouni. *Large Deviations Techniques and Applications*. Jones and Bartlett Publishers, 1993.

[EG02]   J. H. Elder and R. M. Goldberg. Ecological statistics of Gestalt laws for the perceptual organization of contours. *Journal of Vision*, 2(4):324–353, 2002.

[FB81]   M.A. Fischler and R.C. Bolles. Random sample consensus: A paradigm for model fitting with applications to image analysis and automated cartography. *Communications of the ACM*, 24:381–395, 1981.

[FCF+95]   S.D. Forman, J.D. Cohen, M. Fitzgerald, W.F. Eddy, M.A. Mintum, and D.C. Noll.
             Improved assessment of significant activation in functional magnetic resonance
             imaging (fMRI): Use of a cluster-size threshold. *Magnetic Resonance in Medecine*,
             33:636–647, 1995.

[Fel68]   W. Feller. *An introduction to probability theory and its applications*, volume 1.
             Wiley, New York, 3rd edition, 1968.

[Fer06]   F. Fernandez. Mejora al detector de alineamientos. Proyecto final del curso teoria
             computacional de la gestalt. Technical report, Facultad de Ingenieria, Montevideo,
             February 2006.

[FFLF91]   K.J. Friston, C.D. Frith, P.F. Liddle, and R.S.J. Frackowiak. Comparing functional
             ("PET") images: The assessment of significant change. *Journal of Cerebral Blood
             Flow & Metabolism*, 11:690–699, 1991.

[FG01]   F. Fleuret and D. Geman. Coarse-to-fine face detection. *International Journal of
             Computer Vision*, 41:85–107, 2001.

[FL90]   P. Fua and Y.G. Leclerc. Model driven edge detection. *Machine Vision and Appli-
             cations*, 3:45–56, 1990.

[FL01]   O. Faugeras and Q.-T. Luong. *The Geometry of Multiple Images*. MIT Press,
             Cambridge, MA, 2001.

[FLT87]   O. Faugeras, F. Lustman, and G. Toscani. Motion and structure from motion from
             point and line matches. In *International Conference on Computer Vision*, pages
             25–34, 1987.

[FMM98]   J. Froment, S. Masnou, and J.-M. Morel. La géométrie des images naturelles et ses
             algorithmes. Technical Report 98-9, PRISME, Paris Descartes University, 1998.

[GG84]   S. Geman and D. Geman. Stochastic relaxation, gibbs distributions and the
             bayesian restoration of images. *IEEE Transactions on Pattern Analysis and Ma-
             chine Intelligence*, 6:721–741, 1984.

[GK86]   G. Gerig and F. Klein. Fast contour identification through efficient Hough trans-
             form and simplified interpretation strategy. In *Proceedings of the 8th International
             Conference on Pattern Recognition*, volume 1, pages 498–500, Paris, 1986.

[GM96]   G. Guy and G. Medioni. Inferring global perceptual contours from local features.
             International Journal of Computer Vision, 20(1):113–133, 1996.

[GM06]   B. Grosjean and L. Moisan. A-contrario detectability of spots in textured back-
             grounds. Technical Report 2006-12, MAP5, Paris Descartes University, 2006.

[Gom71]   E.H. Gombrich. *The Story of the Art*. Phaidon, London, 1971.

[GPSG01]   W.S. Geisler, J.S. Perry, B.J. Super, and D.P. Gallogly. Edge co-occurrence in
             natural images predicts contour grouping performance. *Vision Research*, 41:711–
             724, 2001.

[Gre80]   U. Grenander. *Abstract Inference*. Wiley, New York, 1980.

[Gre93]   U. Grenander. *General Pattern Theory*. Oxford University Press, 1993.

[Gri99]   G.R. Grimmett. *Percolation*. Springer, New York, 1999.

[GS01]   G.R. Grimmett and D.R. Stirzaker. *Probability and Random Processes*. Oxford
             University Press, New York, 3rd edition, 2001.

[GZW03]   C. Guo, S.-C. Zhu, and Y. N. Wu. Towards a mathematical theory of primal sketch
             and sketchability. In *Proceedings of the Ninth IEEE International Conference on
             Computer Vision (ICCV)*. IEEE Computer Society, New York, 2003.

[Hal88]   P. Hall. *Introduction to the Theory of Coverage Processes*. Wiley Series in Proba-
             bility and Mathematical Statistics: Probability and Mathematical Statistics. Wiley,
             New York, 1988.

[Har84]   R. Haralick. Digital step edges from zero crossing of second derivatives. *IEEE
             Transactions on Pattern Analysis and Machine Intelligence*, 6(1):58–68, 1984.

[Har97a]   R.I. Hartley. Kruppa's equations derived from the fundamental matrix. *IEEE Trans-
             actions on Pattern Analysis and Machine Intelligence*, 19(2):133–135, 1997.

[Har97b]   R.I. Hartley. Self-calibration of stationary cameras. *International Journal of
             Computer Vision*, 22(1):5–24, 1997.

[Her20]   E. Hering. *Grundzüge der Lehre vom Lichtsinn*. Springer-Verlag, Berlin, 1920.

[HGAB03]   N.G. Hatsopoulos, S. Geman, A. Amarasingham, and E. Bienenstock. At what time scale does the nervous system operate? *Neurocomputing*, 52-54:25–29, 2003.

[HM99]   G. Huang and D. Mumford. Statistics of natural images and models. In *International Conference on Computer Vision and Pattern Recognition*, pages 541–547, 1999.

[Hoe63]   W. Hoeffding. Probability inequalities for sum of bounded random variables. *Journal of the American Statistical Association*, 58:13–30, 1963.

[Hor87]   B.K. Horn. *Robot Vision*. MIT Press, Cambridge, MA, 1987.

[HS05]   M. Heiler and C. Schnörr. Natural image statistics for natural image segmentation. *International Journal of Computer Vision*, 63(1):5–19, 2005.

[HT87]   Y. Hochberg and A. C. Tamhane. *Multiple comparison procedures*. Wiley, New York, 1987.

[HZ05]   F. Han and S.-C. Zhu. Bottom-up/top-down image parsing by attribute graph grammar. In *Proceedings of the Tenth IEEE International Conference on Computer Vision (ICCV)*, pages 1778–1785. IEEE Computer Society, New York, 2005.

[IGC05]   L. Igual, L. Garrido, and V. Caselles. A contrast invariant approach to motion estimation. validation and motion segmentation. *Journal of Computer Vision and Image Understanding*, 2005.

[Igu06]   L. Igual. *Image Segmentation and Compression Using the Tree of Shapes of an Image. Motion Estimation*. PhD thesis, Universitat Pompeu Fabra, Barcelona, January 2006.

[JR95]   S. Jensen and L. Rudin. Measure: an interactive tool for accurate forensic photo/video grammetry. In *Investigative & Trial Image Processing Conference, SPIE*, volume 2567, San Diego, CA, 1995.

[Kan79]   G. Kanizsa. *Organization in Vision*. Holt, Rinehart & Winston, New York, 1979.

[Kan91]   G. Kanizsa. *Vedere e pensare*. Il Mulino, Bologna, 1991.

[Kan97]   G. Kanizsa. *Grammatica del Vedere/La Grammaire du Voir*. Il Mulino, Bologna/ Éditions Diderot, Arts et Sciences, 1980 / 1997.

[Kay98]   M. Kay. *Fundamentals of Statistical Signal Processing*. Volume II, Detection Theory. Prentice Hall, 1998.

[KB01]   R. Kimmel and A. Bruckstein. Regularized laplacian zero crossings as optimal edge integrators. In *Proceedings of Image and Vision Computing, IVCNZ01*, New Zealand, 2001.

[KB02]   R. Kimmel and A. Bruckstein. On edge detection, edge integration and geometric active contours. In *Proceedings of Int. Symposium on Mathematical Morphology, ISMM 2002*, Sydney, New South Wales, Australia, 2002.

[KDP05]   G. Koepfler, F. Dibos, and S. Pelletier. Real-time segmentation of moving objects in a video sequence by a contrario detection. In *International Conference on Image Processing*, September 2005.

[KEB91]   N. Kiryati, Y. Eldar, and A.M. Bruckstein. A probabilistic Hough transform. *Pattern Recognition*, 24(4):303–316, 1991.

[Kof35]   K. Koffka. *Principles of Gestalt Psychology*. New York : Harcourt, Brace and Company, 1935.

[KSW85]   J.N. Kapur, P.K. Sahoo, and A.K.C Wong. A new method for gray-level picture thresholding using the entropy of the histogram. *Computer Vision, Graphics and Image Processing*, 29:273–285, 1985.

[KW02]   N. Krüger and F. Wörgötter. Multi-modal estimation of collinearity and parallelism in natural image sequences. *Network: Computation in Neural Systems*, 13(4):553–576, 2002.

[KWT87]   M. Kass, A. Witkin, and D. Terzopoulos. Snakes: active contour models. In *International Conference on Computer Vision*, pages 259–268, 1987.

[Law96]   G. Lawler. *Intersection of Random Walks*. Birkhäuser Boston, 1996.

[LCZ99]   D. Liebowitz, A. Criminisi, and A. Zisserman. Creating architectural models from images. *EuroGraphics*, 18(3), 1999.

[Lec89]   Y. Leclerc. Constructing simple stable descriptions for image partitioning. *International Journal of Computer Vision*, 3:73–102, 1989.

[LF96]   Q.-T. Luong and O. Faugeras. The fundamental matrix: Theory, algorithms and stability analysis. *International Journal of Computer Vision*, 17(1):43–76, 1996.

[LH81]   H. Longuet-Higgins. A computer algorithm for reconstructing a scene from two projections. *Nature*, 293:133–135, 1981.

[Lin97]   M. Lindenbaum. An integrated model for evaluating the amount of data required for reliable recognition. *IEEE Transactions on Pattern Analysis and Machine Intelligence*, 19(11):1251–1264, 1997.

[Lit69]   J. Littlewood. On the probability in the tail of a binomial distribution. *Advances in Applied Probabilities*, 1:43–72, 1969.

[LKB87]   G.E. Legge, D. Kersten, and A.E. Burgess. Contrast discrimination in noise. *Journal of the Optical Society of America A*, 4:391–404, 1987.

[LM03]   J.L. Lisani and J.M. Morel. Detection of major changes in satellite images. In *International Conference on Image Processing*, 2003.

[LMLK94]   E. Lutton, H. Maître, and J. Lopez-Krahe. Contribution to the determination of vanishing points using Hough transform. *IEEE Transactions on Pattern Analysis and Machine Intelligence*, 16(4):430–438, 1994.

[LMR01]   J.-L. Lisani, P. Monasse, and L. Rudin. Fast shape extraction and applications. Technical Report 2001-16, CMLA, ENS Cachan, 2001.

[Low85]   D. Lowe. *Perceptual Organization and Visual Recognition*. Kluwer Academic Publishers, Amsterdam, 1985.

[Mai85]   H. Maitre. Un panorama de la transformation de Hough. *Traitement du signal*, 2(4), 1985.

[Mar72]   A. Martelli. Edge detection using heuristic search methods. *Comparative Graphics Image Processing*, 1:169–182, 1972.

[Mar82]   D. Marr. *Vision*. Freeman and Co., San Francisco, 1982.

[Mee96]   R. Meester. *Continuum Percolation*. Cambridge University Press, 1996.

[Met75]   W. Metzger. *Gesetze des Sehens*. Waldemar Kramer, 1975.

[MG00]   P. Monasse and F. Guichard. Fast computation of a contrast-invariant image representation. *IEEE Transactions on Image Processing*, 9(5):860–872, 2000.

[MH80]   D. Marr and E. Hildreth. Theory of edge detection. *Proceedings Royal Society London*, B 207:187–217, 1980.

[Moi01]   L. Moisan. Asymptotic estimates and inequalities for the tail of the binomial distribution. Unpublished, 2001.

[Mon71]   U. Montanari. On the optimal detection of curves in noisy pictures. *CACM*, 14(5):335–345, 1971.

[Mon00]   P. Monasse. *Représentation morphologique d'images numériques et aplication au recalage d'images*. PhD thesis, Paris Dauphine University, 2000.

[MS85]   D. Mumford and J. Shah. Boundary detection by minimizing functionals. In *Proceedings IEEE Conference on Computer Vision and Pattern Recognition*, San Francisco, 1985.

[MS89]   D. Mumford and J. Shah. Optimal approximations by piecewise smooth functions and associated variational problems. *Communications on Pure and Applied Mathematics*, 42(4):577–685, 1989.

[MS93]   N. Madras and G. Slade. *The Self-avoiding Walk*. Probability and Its Applications. Birkhäuser, 1993.

[MS94]   J.-M. Morel and S. Solimini. *Variational Methods in Image Segmentation*. Birkhäuser, Boston, 1994.

[MS04]   L. Moisan and B. Stival. A probabilistic criterion to detect rigid point matches between two images and estimate the fundamental matrix. *International Journal of Computer Vision*, 57(3):201–218, 2004.

[MSC$^+$06a] P. Musé, F. Sur, F. Cao, Y. Gousseau, and J.-M. Morel. An a contrario decision method for shape element recognition. *International Journal of Computer Vision*, 69(3):295–315, 2006.

[MSC+06b]   P. Musé, F. Sur, F. Cao, Y. Gousseau, and J.-M. Morel. Shape recognition based on an a contrario methodology. In *Statistics and Analysis of Shapes*. H. Krim and A. Yezzi eds, Birkhaüser, Boston, 2006.

[MSCG03]   P. Musé, F. Sur, F. Cao, and Y. Gousseau. Unsupervised thresholds for shape matching. In *International Conference on Image Processing*, 2003.

[Mus04]   P. Musé. *Sur la définition et la reconnaissance de formes planes dans les images numériques*. PhD thesis, École Normale Supérieure de Cachan, October 2004.

[MVM05]   L. Moisan, E. Villéger, and J.-M. Morel. Detection of constant width in images. Unpublished, 2005.

[NMS93]   N. Nitzberg, D. Mumford, and T. Shiota. *Filtering, Segmentation and Depth*. Springer-Verlag, New-York, 1993.

[OF96]   B.A. Olshausen and D.J. Field. Emergence of simple-cell receptive field properties by learning a sparse code for natural images. *Nature*, 381(6583):607–609, 1996.

[OG01]   J. Oliensis and Y. Genc. Fast and accurate algorithms for projective multi-image structure from motion. *IEEE Transactions on Pattern Analysis and Machine Intelligence*, 23(6):546–559, 2001.

[Oka58]   M. Okamoto. Some inequalities relating to the partial sum of binomial probabilities. *Annals of the Institute of Statistical Mathematics*, 10:29–35, 1958.

[OP03]   S. Osher and N. Paragios, editors. *Geometric Level Set Methods in Imaging, Vision and Graphics*. Springer-Verlag, New York, 2003.

[OS88]   S. Osher and J.A. Sethian. Fronts propagating with curvature-dependent speed: Algorithms based on hamilton-jacobi formulations. *Journal of Computational Physics*, 79(1):12–49, 1988.

[Pav86]   T. Pavlidis. A critical survey of image analysis methods. In *IEEE Proceedings of the 8th International Conference on Pattern Recognition*, pages 502–511, Paris, 1986.

[PFM03]   D.G. Pelli, B. Farell, and D.C. Moore. The remarkable inefficiency of word recognition. *Nature*, 423:752–756, 2003.

[PHB99]   J. Puzicha, T. Hofmann, and J.M. Buhmann. Histogram clustering for unsupervised image segmentation. *International Conference on Computer Vision and Pattern Recognition*, pages 602–608, 1999.

[PIK94]   J. Princen, J. Illingowrth, and J. Kittler. Hypothesis testing: A framework for analyzing and optimizing hough transform performance. *IEEE Transactions on Pattern Analysis and Machine Intelligence*, 16(4):329–341, 1994.

[Pre06]   J. Preciozzi. Report of merging algorithms, proyecto de master. Technical report, Facultad de Ingenieria, Montevideo, April 2006.

[Pro61]   Y. Prohorov. Asymptotic behavior of the binomial distribution. *Selected Translations in Math. Stat. and Prob., AMS*, 1:87–95, 1961.

[PTVF88]   W.H. Press, S.A. Teukolsky, W.T. Vetterling, and B.P. Flannery. *Numerical Recipes in C*. Cambridge University Press, 1988.

[Pun81]   T. Pun. Entropic thresholding, a new approach. *Computer Graphics and Image Processing*, 16:210–239, 1981.

[PY03]   M. D. Penrose and J. E. Yukich. Weak laws of large numbers in geometric probability. *The Annals of Applied Probability*, 13(1):277–303, 2003.

[PZ89]   P. Parent and S.W. Zucker. Trace inference, curvature consistency and curve detection. *IEEE Transactions on Pattern Analysis and Machine Intelligence*, 2(8), 1989.

[QT97]   J. Quintanilla and S. Torquato. Clustering in a continuum percolation model. *Advances in Applied Probability*, 29(2):327–336, 1997.

[Ris83]   J. Rissanen. A universal prior for integers and estimation by minimum description length. *Annals of Statistics*, 11(2):416–431, 1983.

[Ris89]   J. Rissanen. *Stochastic Complexity in Statistical Inquiry*. World Scientific Press, Singapore, 1989.

[RMHM05]   A. Robin, L. Moisan, and S. Le Hégarat-Mascle. Automatic land-cover change detection from coarse resolution images using an a contrario approach. Technical Report 2005-3, MAP5, Paris Descartes University, 2005.

[ROF92]   L. Rudin, S. Osher, and E. Fatemi. Nonlinear total variation based noise removal algorithms. *Physica D*, 60(1-4):259–268, 1992.

[Rot00]   C. Rother. A new approach for vanishing point detection in architectural environments. In *British Machine Vision Conference*, 2000.

[RT71]   A. Rosenfeld and M. Thurston. Edge and curve detection for visual scene analysis. *IEEE Transactions on Computing*, 20:562–569, 1971.

[RT02]   R. Roy and H. Tanemura. Critical intensities of Boolean models with different underlying convex shapes. *Advances in Applied Probability*, 34(1):48–57, 2002.

[Rub15]   E. Rubin. *Visuell wahrgenommene Figuren*. (Transl. of 1915 original publication into German), Kopenhagen: Gyldendal, 1915.

[SA94]   D. Stauffer and A. Aharony. *Introduction to Percolation Theory*. Taylor and Francis, 1994.

[SA96]   E.P. Simoncelli and E.H. Adelson. Noise removal via bayesian wavelet coring. In *Proceedings of the 3rd International Conference on Image Processing (ICIP)*, pages I: 379–382, Lausanne, 1996.

[San76]   L. Santaló. Integral geometry and geometric probability. In Gian-Carlo Rota, editor, *Encyclopedia of Mathematics and its Applications*, volume 1. Addison-Wesley, 1976.

[Sap90]   G. Saporta. *Probabilités, analyse des données et statistique*. Editions Technip, 1990.

[SAP01]   J. Salvi, X. Armangué, and J. Pagès. A survey addressing the fundamental matrix estimation problem. In *International Conference on Image Processing*, 2001.

[Ser82]   J. Serra. *Image Analysis and Mathematical Morphology*. Academic Press, New York, 1982.

[Ser88]   J. Serra. *Image Analysis and Mathematical Morphology, Part. II: Theoretical Advances*. Academic Press, 1988.

[Sha48]   C.E. Shannon. A mathematical theory of communication. *Bell System Technical Journal*, 27:379–423 and 623–656, 1948.

[Shu99]   J.A. Shufelt. Performance evaluation and analysis of vanishing point detection techniques. *IEEE Transactions on Pattern Analysis and Machine Intelligence*, 21(3):282–288, 1999.

[SKM87]   D. Stoyan, W.S. Kendall, and J. Mecke. *Stochastic geometry and its applications*. Wiley Series in Probability and Mathematical Statistics: Applied Probability and Statistics. Wiley, 1987.

[Slu77]   E. Slud. Distribution inequalities for the binomial law. *Annals of Probability*, 5:404–412, 1977.

[Sma96]   C. G. Small. *The Statistical Theory of Shape*. Springer-Verlag, New York, 1996.

[SS94]   D. Stoyan and H. Stoyan. *Fractals, random shapes and point fields*. Wiley Series in Probability and Mathematical Statistics: Applied Probability and Statistics. Wiley, 1994.

[Ste95]   C.V. Stewart. MINPRAN: A new robust estimator for computer vision. *IEEE Transactions on Pattern Analysis and Machine Intelligence*, 17:925–938, 1995.

[Ste02]   J. M. Steele. Minimal spanning trees for graphs with random edge lengths. In *Mathematics and Computer Science, II*, Trends Math., pages 223–245. Birkhäuser, Basel, 2002.

[SU88]   A. Sha'Ashua and S. Ullman. Structural saliency: The detection of globally salient structures using a locally connected network. In *International Conference on Computer Vision*, pages 321–327, 1988.

[Sur04]   F. Sur. *Décision a contrario pour la reconnaissance de formes*. PhD thesis, École Normale Supérieure de Cachan, October 2004.

[SYK96]   D. Shaked, O. Yaron, and N. Kiryati. Deriving stopping rules for the probabilistic Hough transform by sequential analysis. *Journal of Computer Vision and Image Understanding*, 63(3):512–526, 1996.

[SZ00]   F. Schaffalitzky and A. Zisserman. Planar grouping for automatic detection of vanishing lines and points. *Image and Vision Computing*, 18(9):647–658, 2000.

[Tal95]   M. Talagrand. The missing factor in Hoeffding's inequalities. *Annales Institut Henri Poincaré*, 31(4):698–702, 1995.

[TC92]   D.-M. Tsai and Y.-H. Chen. A fast histogram-clustering approach for multi-level thresholding. *Pattern Recognition Letters*, 13:245–252, 1992.

[TM97]   P.H.S. Torr and D.W. Murray. The development and comparison of robust methods for estimating the fundamental matrix. *International Journal of Computer Vision*, 24(3):271–300, 1997.

[TML01]   C.K. Tang, G. Medioni, and M.S. Lee. N-dimensional tensor voting, and application to epipolar geometry estimation. *IEEE Transactions on Pattern Analysis and Machine Intelligence*, 23(8):829–844, 2001.

[TPG97]   T. Tuytelaars, M. Proesmans, and L. Van Gool. The cascaded Hough transform. In *International Conference on Image Processing*, volume 2, pages 736–739, 1997.

[Tre68]   H.L. Van Trees. *Detection, Estimation and Modulation Theory*, volume 1. Wiley, New York, 1968.

[TZM95]   P.H.S. Torr, A. Zisserman, and D.W. Murray. Motion clustering using the trilinear constraint over three views. In *Workshop on Geometrical Modeling and Invariants for Computer Vision*. Xidian University Press, 1995.

[VC02]   L.A. Vese and T.F. Chan. A multiphase level set framework for image segmentation using the mumford and shah model. *International Journal of Computer Vision*, 50(3):271–293, 2002.

[VCB04]   T. Veit, F. Cao, and P. Bouthemy. Probabilistic parameter-free motion detection. In *International Conference on Computer Vision and Pattern Recognition*, volume I, pages 715–721. IEEE, 2004.

[VCB05]   T. Veit, F. Cao, and P. Bouthemy. A maximality principle applied to a contrario motion detection. In *International Conference on Image Processing*, 2005.

[vH99]   H. von Helmholtz. *Treatise on Physiological Optics*. Thoemmes Press, 1999.

[WB91]   D.M. Wuescher and K.L. Boyer. Robust contour decomposition using constant curvature criterion. *IEEE Transactions on Pattern Analysis and Machine Intelligence*, 13(1):41–51, 1991.

[Wer23]   M. Wertheimer. Untersuchungen zur lehre der gestalt, II. *Psychologische Forschung*, 4:301–350, 1923.

[WT83]   A.P. Witkin and J. Tenenbaum. On the role of structure in vision. In *Human and Machine Vision*, pages 481–543. A. Rosenfeld ed., Academic Press, New York, 1983.

[Wu00]   X. Y. Wu. Self-containing property of Euclidean minimal spanning trees on infinite random points. *Acta Mathematica Sinica*, 43(1):107–116, 2000.

[XBA03]   N. Xu, R. Bansal, and N. Ahuja. Object segmentation using graph cuts based active contours. In *Proceedings of the Int. Conf. on Computer Vision and Pattern Recognition (CVPR)*, pages II: 46–53, 2003.

[Yuk00]   J. E. Yukich. Asymptotics for weighted minimal spanning trees on random points. *Stochastic Processes and their Applications*, 85(1):123–138, 2000.

[ZDFL94]   Z. Zhang, R. Deriche, O. Faugeras, and Q.-T. Luong. A robust technique for matching two uncalibrated images through the recovery of the unknown epipolar geometry. *AI Journal*, 78:87–119, 1994.

[ZDFL01]   Z. Zhang, R. Deriche, O. Faugeras, and Q.T. Luong. Estimating the fundamental matrix by transforming image points in projective space. *Journal of Computer Vision and Image Understanding*, 82:174–180, 2001.

[Zha98]   Z. Zhang. Determining the epipolar geometry and its uncertainty: A review. *International Journal of Computer Vision*, 27(2):161–195, 1998.

[Zhu99]   S.C. Zhu. Embedding gestalt laws in markov random fields. *IEEE Transactions on Pattern Analysis and Machine Intelligence*, 21(11):1170–1187, 1999.

[Zuc76]   S.W. Zucker. Region growing: Childhood and adolescence (survey). *Computer Graphics and Image Processing*, 5:382–399, 1976.

Index

Interdisciplinary Applied Mathematics